Synthesis Lectures on Mathematics & Statistics

Series Editor

Steven G. Krantz, Department of Mathematics, Washington University, Saint Louis, MO, USA

This series includes titles in applied mathematics and statistics for cross-disciplinary STEM professionals, educators, researchers, and students. The series focuses on new and traditional techniques to develop mathematical knowledge and skills, an understanding of core mathematical reasoning, and the ability to utilize data in specific applications.

Liliana Blanco-Castañeda ·
Viswanathan Arunachalam

Applied Stochastic Modeling

 Springer

Liliana Blanco-Castañeda
Universidad Nacional de Colombia
Bogotá, Colombia

Viswanathan Arunachalam
Universidad Nacional de Colombia
Bogotá, Colombia

ISSN 1938-1743   ISSN 1938-1751 (electronic)
Synthesis Lectures on Mathematics & Statistics
ISBN 978-3-031-31281-6   ISBN 978-3-031-31282-3 (eBook)
https://doi.org/10.1007/978-3-031-31282-3

This Springer imprint is published by the registered company Springer Nature Switzerland AG
The registered company address is: Gewerbestrasse 11, 6330 Cham, Switzerland

Preface

This book is designed as a reference text for students and researchers who need to consult stochastic models in their professional work and are unfamiliar with the mathematical and statistical theory required to understand these methods. The most relevant concepts and results for the development of the examples are presented, omitting rigorous mathematical proofs and giving only the guidelines of those fundamental in constructing the models.

The book is divided into five chapters. Chapter 1 presents a compilation of the discrete-time Markov chain and the most relevant concepts and theorems. Chapter 2 deals with the Poisson process and some important properties. Chapter 3 is devoted to studying the continuous-time Markov chain and its applications, with particular emphasis on the birth and death process. Chapter 4 deals with Branching processes, particularly the Galton-Watson process with two types of individuals and is presented as an example to describe the evolution of the SARS-CoV2 virus in Bogota. Chapter 5 presents the theory corresponding to a hidden Markov model developed for the behavior of the horizontal displacements of the behaviors of two animals from their observed trajectories in order to identify hidden behavioral states and determine the preferences of habitat.

We thank our students for their assistance in typesetting and preparing programming codes, which has served as the platform for this project. We thank our Editors, Ms. Susanne Steitz-Filler and Ms. Melanie Rowen, Springer Nature, for their advice and technical support.

Bogotá, Colombia   Liliana Blanco-Castañeda
February 2023   Viswanathan Arunachalam

Contents

Discrete-Time Markov Chain

Markov chains are named after the Russian mathematician Andrei Andreyevich Markov (1856–1922) who introduced them in his work "Extension of the law of large numbers to dependent quantities", published in 1906, in which he developed the concept of the law of large numbers and the central limit theorem for sequences of dependent random variables [1]. As a disciple of the Russian mathematician Patnufy Chebyschev (1821–1894), he made great contributions to probability theory, number theory, and analysis. He worked as a professor at the University of Saint Petersburg since 1886, from where he retired in 1905, although he continued teaching until the end of his life.

Markov developed his theory of chains from a completely theoretical point of view, he also applied these ideas to chains of two states, vowels, and consonants, in some literary texts of the Russian poet Aleksandr Pushkin (1799–1837). Markov analyzed the sequences of vowels and consonants in Pushkin's verse work "Eugene Onegin", concluding that the letters are not arranged independently in the poem but that the placement of each letter depends on the previous letter.

Markov lived through a period of great political activity in Russia and became actively involved. In the year 1902, the Russian novelist, Maxim Gorky was elected to the Russian Academy of Sciences in 1902, but the direct order of the Tsar canceled his election. Markov protested and refused the honors he was awarded the following year. Later, when the interior ministry ordered university professors to report any anti-government activity by their students, he objected, claiming that he was a professor of probability and not a policeman [2]. Currently, Markov chains are used to find the author of a text [3] and in web search systems such as Google [4].

L. Blanco-Castañeda and V. Arunachalam, *Applied Stochastic Modeling*, Synthesis Lectures on Mathematics & Statistics, https://doi.org/10.1007/978-3-031-31282-3_1

1.1   Introduction to Stochastic Processes

Definition 1.1 A *stochastic process* is a family or a collection of random variables $X = \{X_t, t \in T\}$ defined on a common probability space $(\Omega, \Im, P)$ with taking values in a measurable space $(S, \mathcal{S})$, called the *state space*. The set of parameters T is called the *parameter space* of the stochastic process, which is usually a subset of $\mathbb{R}$.

The mapping defined for each fixed $\omega \in \Omega$, the function $t \rightarrow X_t(\omega), t \in \mathbb{R}$, is called *sample path* or a *realization* of the stochastic process X. The process path associated with ω provides a mathematical model for a random experiment whose outcome can be observed continuously in time.

The set of possible values of the indexing parameter which can be either discrete or continuous. For our convenience, when the indexing parameter is discrete, the family is represented by $\{X_n, n = 0, 1, 2 \ldots\}$. In case of continuous time both $\{X_t, t \in T\}$ and $\{X(t), t \in T\}$ are used. If the state space and the parameter space of a stochastic process are discrete, then the process is called stochastic sequence, and often referred as a chain.

Stochastic processes can be classified, in general, into the following four types of processes:

1. Discrete time, discrete state space (DTDS).
2. Discrete time, continuous state space (DTCS).
3. Continuous time, discrete state space (CTDS).
4. Continuous time, continuous state space (CTCS).

Definition 1.2 Let $\{X_t; t \in T\}$ be a stochastic process and $\{t_1, t_2, \ldots, t_n\} \subset T$ where $t_1 < t_2 < \cdots < t_n$. The function

$$F_{t_1 \ldots t_n}(x_1, \ldots, x_n) := P(X_{t_1} \leq x_1, \ldots, X_{t_n} \leq x_n)$$

is called *the finite-dimensional distribution of the process*.

Definition 1.3 If, for all $t_0, t_1, t_2, \ldots, t_n$ such that $t_0 < t_1 < t_2 < \cdots < t_n$, the random variables $X_{t_0}, X_{t_1} - X_{t_0}, X_{t_2} - X_{t_1}, \ldots, X_{t_n} - X_{t_{n-1}}$ are independent, then the process $\{X_t; t \in T\}$ is said to be a process with independent increments.

Definition 1.4 A stochastic process $\{X_t; t \in T\}$ is said to have *stationary increments* if $X_{t_2+s} - X_{t_1+s}$ has the same distribution as $X_{t_2} - X_{t_1}$ for all choices of $t_1, t_2 \in T$ and $s > 0$.

Definition 1.5 If for all $t_1, t_2, \ldots, t_n$ the joint distributions of the vector random variables

$$(X(t_1), X(t_2), \ldots, X(t_n)) \text{ and } (X(t_1 + h), X(t_2 + h) \ldots, X(t_n + h))$$

are the same for all $h > 0$, then the stochastic process $\{X_t; t \in T\}$ is said to be a stationary stochastic process of order n (or simply a stationary process). The stochastic process $\{X_t; t \in T\}$ is said to be a strong stationary stochastic process or strictly stationary process if the above property is satisfied for all n.

Definition 1.6 A stochastic process $\{X_t; t \in T\}$ is called a second-order process if $E\left(X_t^2\right) < \infty$ for all $t \in T$.

Example 1.1 Let Z_1 and Z_2 be independent normally distributed random variables, each having mean 0 and variance σ^2. Let $\lambda \in \mathbb{R}$ and

$$X_t = Z_1 \cos(\lambda t) + Z_2 \sin(\lambda t), \quad t \in \mathbb{R}.$$

$\{X_t; t \in T\}$ is a second-order stationary process.   ▲

Definition 1.7 A second-order stochastic process $\{X_t; t \in T\}$ is called *covariance stationary* or *weakly stationary* if its mean function $m(t) = E[X_t]$ is independent of t and its covariance function $Cov(X_s, X_t)$ depends only on the difference $|t - s|$ for all $s, t \in T$. That is:
$$Cov(X_s, X_t) = f(|t - s|).$$

Definition 1.8 A stochastic process that is not stationary in any sense is called an evolutionary stochastic process.

Definition 1.9 A stochastic process $\{X_t; t \in T\}$ is a *Gaussian process* if the random vectors $(X(t_1), X(t_2), \ldots, X(t_n))$ have a joint Normal distribution for all $(t_1, t_2, \ldots, t_n)$ and $t_1 < t_2 < \cdots < t_n$.

Definition 1.10 Let $\{X_t; t \geq 0\}$ be a stochastic process defined over a probability space $(\Omega, \Im, P)$ and with state space $(\mathbb{R}, \mathcal{B})$. We say that the stochastic process $\{X_t; t \geq 0\}$ is called a Markov process if for any $0 \leq t_1 < t_2 < \cdots < t_n$ and for any states $B, B_1, B_2, \ldots, B_{n-1} \in \mathcal{B}$:

$$P\left(X_{t_n} \in B \mid X_{t_1} \in B_1, \ldots, X_{t_{n-1}} \in B_{n-1}\right) = P\left(X_{t_n} \in B \mid X_{t_{n-1}} \in B_{n-1}\right). \quad (1.1)$$

The above condition (1.1) is called the Markov property, and has the following implications: Any stochastic process with independent increments is a Markov process. Also, the Markov process is such that, given the value of X_s, for $t > s$, the distribution of X_t does not depend on the values of X_u, for $u < s$.

1.2   Discrete-Time Markov Chain

The Markov chain is defined as a sequence of random variables taking a finite or countable set of values and characterized by the Markov property. This section discusses the most important properties of the discrete-time Markov chain (for more details see [5, 6]).

Definition 1.11 The stochastic process $\{X_n; n \in \mathbb{N}\}$ with $n = 0, 1, \ldots$ is called a *discrete-time Markov chain* if for all for all $t_0 < t_1 < \cdots < t_{n+1}$ with $t_i \in T$ and $i, j, i_0, i_1, \ldots, i_{n-2} \in S$ We have

$$P(X_n = j \mid X_{n-1} = i, X_{n-2} = i_{n-2}, \ldots, X_0 = i_0) = P(X_n = j \mid X_{n-1} = i) \quad (1.2)$$

with

$$P(X_0 = i_0, \ldots, X_{n-1} = i) > 0.$$

Here the future state $X_n = j$ of the Markov chain depends only on the present state $X_{n-1} = i$, but not on the past "$X_{n-2}, X_{n-3}, \ldots, X_0$".

Let $\{X_n; n \in \mathbb{N}\}$ be a discrete-time Markov chain. If $X_0 = i_0$, then the chain is said to have started in the state i_0. If $X_n = i_n$ then the chain is said to be at time n in state i_n. The sequence of states $i_0, i_1, \ldots, i_n$ is said to be the complete history of the chain up to the time n, if $X_0 - i_0$, $X_1 = i_1, \ldots, X_n = i_n$.

Theorem 1.1 *The stochastic process $\{X_n; n \in \mathbb{N}\}$ with set of states S is a Markov chain, if and only if, for any finite sequence of natural numbers $0 \leq n_0 < n_1 < \cdots < n_k$ and for any choice $i_{n_0}, i_{n_1}, \ldots, i_{n_k} \in S$ it is satisfied that:*

$$P\left(X_{n_k+m} = j \mid X_{n_k} = i_{n_k}, \ldots, X_{n_0} = i_{n_0}\right) = P\left(X_{n_k+m} = j \mid X_{n_k} = i_{n_k}\right) \quad (1.3)$$

for any integer $m > 1$.

Definition 1.12 Let $\{X_n; n \in \mathbb{N}\}$ be a Markov chain with discrete-time parameter. The probabilities

$$p_{ij} := P(X_{n+1} = j \mid X_n = i) \quad (1.4)$$

with $i, j \in S$ are called *transition probabilities*. The matrix form of the transition probability is written as

$$\mathbf{P} = (p_{ij}) = \begin{pmatrix} p_{00} & p_{01} & p_{02} & \cdots \\ p_{10} & p_{11} & p_{12} & \cdots \\ p_{20} & p_{21} & p_{22} & \cdots \\ \vdots & \vdots & \vdots & \ddots \end{pmatrix}$$

is called the *transition probabilities matrix* or *stochastic matrix*, and satisfies the following:

$$p_{ij} \geq 0 \text{ for all } i, j \in S$$

$$\sum_j p_{ij} = 1 \text{ for all } i \in S.$$

Remark 1.1 • A Markov chain $\{X_n; n \geq 0\}$ is called *homogeneous* if the transition probabilities do not depend on time-step n. That, is for $n \in \mathbb{N}$

$$p_{ij} := P(X_1 = j \mid X_0 = i) = P(X_{n+1} = j \mid X_n = i).$$

• The transition probabilities with the initial distribution $\pi_i^{(0)} := P(X_0 = i)$ completely determines the Markov chain. That is, if $\{X_n, n = 0, 1, 2, \dots\}$ is a Markov chain, then for all n and $i_0, \dots, i_n$ the set of states the satisfies the following:

$$P(X_0 = i_0, \dots, X_n = i_n) = \pi_i^{(0)} P(X_1 = i_1 \mid X_0 = i_0) \dots P(X_n = i_n \mid X_{n-1} = i_{n-1}).$$

Example 1.2 Suppose a random experiment is performed where there are only two possible outcomes success or failure, with a probability of success $0 < p < 1$ and probability of failure $q := 1 - p$. Let X_n be the random variable denoting the number of successes in n repetitions of the experiment. The random variable X_n has a binomial distribution of parameters n and p and the sequence $\{X_n; n \geq 1\}$ is a Markov chain with state set $S = \{0, 1, 2, \dots\}$ and transition matrix

$$\mathbf{P} = (p_{i,j})_{i,j \in S}$$

with

$$p_{ij} = \begin{cases} p & \text{if } j = i+1 \\ q & \text{if } \quad j = i \\ 0 & \text{otherwise} \end{cases}$$

Example 1.3 (*Random walk*) Let $(Y_n)_{n \geq 1}$ be a sequence of independent and equally distributed variables with values in $\mathbb{Z}$. The process $\{X_n; n \geq 0\}$ defined by:

$$X_0 := 0$$

$$X_n := \sum_{k=1}^{n} Y_k$$

is a Markov chain with state set $\mathbb{Z}$ and matrix of transition $\mathbf{P} = (p_{i,j})_{i,j \in \mathbb{Z}}$ where $p_{i,j} = P(Y_1 = j - i)$.

Example 1.4 Suppose we have two players A and B at the beginning of the game, player A has a capital of $x \in \mathbb{Z}^+$ dollar and player B a capital of $y \in \mathbb{Z}^+$ dollar. Say $a := x + y$. In each round of the game, either A wins B a dollar with probability p or B wins A a dollar with probability q being $p + q = 1$. The game continues until one of the two players loses his capital, that is, until $X_n = 0$ or $X_n = a$.

Let $X_n :=$ "capital of A after the nth game round." The sequence $(X_n)_{n \in \mathbb{N}}$ is a Markov chain with set of states $S = \{0, 1, 2, \ldots, a\}$ and transition matrix is given by

$$\mathbf{P} = \begin{pmatrix} 1 & 0 & 0 & 0 & \cdots & 0 \\ q & 0 & p & 0 & \cdots & 0 \\ 0 & q & 0 & p & \cdots & 0 \\ \vdots & \vdots & \vdots & \vdots & \cdots & \vdots \\ \vdots & \vdots & \vdots & \vdots & \cdots & \vdots \\ 0 & 0 & 0 & 0 & \cdots & 1 \end{pmatrix}$$

Definition 1.13 Let $\{X_n; n \in \mathbb{N}\}$ be a Markov chain. The transition probability in m steps, $p_{ij}^{(m)}$, is the probability that from the state i, the state reached at state j at the mth step and defined as

$$p_{ij}^{(m)} = P(X_m = j \mid X_0 = i). \tag{1.5}$$

The $p_{ij}^{(m)}$ is stationary, if and only if, for all $n \in \mathbb{N}$

$$p_{ij}^{(m)} = P(X_{n+m} = j \mid X_n = i) = P(X_m = j \mid X_0 = i) \tag{1.6}$$

A Markov chain whose transition probabilities in m steps are all stationary is called a homogeneous Markov chain. The transition matrix for $m-$ transition probabilities is written as

$$\mathbf{p}^m = \left(p_{ij}^m\right)_{i,j \in S} \tag{1.7}$$

Homogeneous Markov chains can be represented by a network in which the vertices indicate the states of the chain, and the arcs indicate the transitions between one state and another. For example, if $\{X_n; n \in \mathbb{N}\}$ is a Markov chain with set of states $S = \{0, 1, 2, 3\}$ with transition probability matrix

$$\mathbf{P} = \begin{pmatrix} \frac{1}{5} & \frac{1}{5} & 0 & \frac{3}{5} \\ 0 & \frac{1}{3} & \frac{2}{3} & 0 \\ \frac{1}{2} & 0 & 0 & \frac{1}{2} \\ \frac{1}{4} & \frac{1}{4} & \frac{1}{4} & \frac{1}{4} \end{pmatrix}.$$

The graphical representation of the state transition is shown in Fig. 1.1.

The following Chapman-Kolmogorov equation gives a method of computing $n-$step transition probabilities.

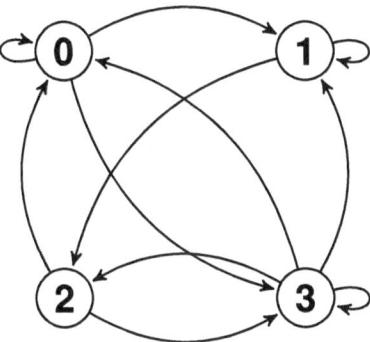

Fig. 1.1 State transition diagram

Proposition 1.1 *If $\{X_n; n \in \mathbb{N}\}$ is a homogeneous Markov chain and if $k < m < n$ then for states $h, i, j \in S$, we have*

$$p_{hj}^n = \sum_{i \in S} p_{ij}^{n-m} p_{hi}^m. \tag{1.8}$$

Remark 1.2 The above proposition which states that the transition matrix in m steps is the mth power of the transition matrix. That is,

$$\mathbf{P}^{(n)} = \mathbf{P}^n \tag{1.9}$$

Example 1.5 A Markov chain $\{X_n; n \geq 1\}$ with set of states $S = \{0, 1\}$ and transition matrix

$$\mathbf{P} = \begin{pmatrix} 1-a & a \\ b & 1-b \end{pmatrix}$$

where a and b are real numbers with $0 < a < 1$ and $0 < b < 1$.

The eigenvalues of the matrix $\mathbf{P}$ are $\lambda_1 = 1$ and $\lambda_2 = 1 - a - b$ and the corresponding eigenvectors are

$$v_1 = \begin{pmatrix} 1 \\ 1 \end{pmatrix} \quad \text{and} \quad v_2 = \begin{pmatrix} -a \\ b \end{pmatrix}$$

Then

$$\mathbf{P} = ADA^{-1}$$

with

$$A = \begin{pmatrix} 1 & -a \\ 1 & b \end{pmatrix}$$

$$D = \begin{pmatrix} 1 & 0 \\ 0 & 1-a-b \end{pmatrix}$$

Since

$$A^{-1} = \frac{1}{a+b} \begin{pmatrix} b & a \\ -1 & 1 \end{pmatrix}$$

we get

$$\mathbf{P}^n = AD^n A^{-1}$$
$$= \frac{1}{a+b} \begin{pmatrix} 1 & -a \\ 1 & b \end{pmatrix} \begin{pmatrix} 1 & 0 \\ 0 & (1-a-b)^n \end{pmatrix} \begin{pmatrix} b & a \\ -1 & 1 \end{pmatrix}$$
$$= \frac{1}{a+b} \begin{pmatrix} b+a(1-a-b)^n & a(1-(1-a-b)^n) \\ b(1-(1-a-b)^n) & a+b(1-a-b)^n \end{pmatrix}.$$

Proposition 1.2 *If $\{X_n; n \in \mathbb{N}\}$ is a homogeneous Markov chain with transition matrix $\mathbf{P}$, set of states S and initial distribution $\pi^{(0)}$ then for all $n \in \mathbb{N}$ and for all $k \in S$ we have*

$$P(X_n = k) = \sum_{j \in S} p_{jk}^n \pi_j^{(0)} \tag{1.10}$$

Definition 1.14 Let $\{X_n; n \in \mathbb{N}\}$ be a homogeneous Markov chain with transition matrix $\mathbf{P}$, set of states S and initial distribution $\pi^{(0)}$. The chain is said to have a limit distribution $\pi = (\pi_0, \pi_1, \dots)$ if for all $j \in S$ is satisfied that

$$\lim_{n \to \infty} P(X_n = j) = \pi_j \tag{1.11}$$

Definition 1.15 Let $\{X_n; n \in \mathbb{N}\}$ be a homogeneous Markov chain with transition matrix $\mathbf{P} = (p_{i,j})_{i,j \in S}$, set of states S and initial distribution $\pi^{(0)}$. The chain is said to have a stationary distribution $\pi = (\pi_0, \pi_1, \dots)$ if for all $j \in S$ is satisfies the following

$$\pi_j = \sum_{i \in S} \pi_i p_{ij}, \tag{1.12}$$

and

$$\sum_{j \in S} \pi_j = 1 \tag{1.13}$$

Remark 1.3 A homogeneous Markov chain $\{X_n; n \in \mathbb{N}\}$ with transition matrix $\mathbf{P}$. Suppose that π is a stationary distribution of the chain, then

$$\sum_{i \in S} \pi_i P_{ij}^2 = \sum_{i \in S} \pi_i \sum_{k \in S} p_{ik} p_{kj}$$
$$= \sum_{k \in S} \left(\sum_{i \in S} \pi_i p_{ik} \right) p_{kj}$$
$$= \sum_{k \in S} \pi_k p_{kj} = \pi_j.$$

Using inductive reasoning, we have shown that

$$\sum_{i \in S} \pi_i p_{ij}^n = \pi_j$$

Therefore, then for each $j \in S$ and each $n \in \mathbb{N}$ is satisfied:

$$P(X_n = j) = \pi_j. \tag{1.14}$$

The distribution of X_n is independent of n. Note that a limit distribution is always stationary. The converse is not always hold.

Classification of States

It is interesting to know the conditions that guarantee the existence and uniqueness of stationary distributions and their interpretation. In order to find answers to these concerns, we briefly introduce the following concepts of classification of states and give some fundamental properties of the Markov chain.

Definition 1.16 Let $\{X_n; n \geq 0\}$ be a Markov chain.

- State j is said to be accessible from state i if for some $n \in \mathbb{N}$ such that $p_{ij}^{(n)} > 0$. In such a case we write $i \mapsto j$.
- State i is said to be accessible from state j if for some $m \in \mathbb{N}$ such that $p_{ji}^{(m)} > 0$. In such a case we write $j \mapsto i$.
- State i is said to be in communication with state j and write $i \longleftrightarrow j$, if $i \mapsto j$ and $j \mapsto i$.

Remark 1.4 1. A subset $\phi \neq A \subseteq S$ is said to be closed, if and only if, for all $i, j \in A$ we have to: $i \longleftrightarrow j$.
2. A closed set may contain one or more states. If a closed set contains only one state then the state is called an absorbing state.
3. Every finite Markov chain contains at least one closed set.
4. A Markov chain is called irreducible if there exists no nonempty closed set other than S itself. If S has a proper closed subset, then it is called reducible chain.

Definition 1.17 Let $\{X_n; n \in \mathbb{N}\}$ be a homogeneous Markov chain. The probability of recurrence in n steps of the state $i \in S$ is defined as follows:

$$f_i^{(n)} = P(X_n = i \mid X_{n-1} \neq i, \ldots, X_1 \neq i, X_0 = i) \tag{1.15}$$

and f_i is defined by:

$$f_i := \sum_{n=1}^{\infty} f_i^{(n)} \qquad (1.16)$$

is called the probability of recurrence of the state i.

Definition 1.18 A state $i \in S$ is called transient if $f_i < 1$ and recurrent if $f_i = 1$.

Definition 1.19 Let $i \in S$ be a recurrent state. The mean return time is defined to state i as follows:

$$m_i := \sum_{n=1}^{\infty} n f_i^{(n)} \qquad (1.17)$$

If $m_i < \infty$ then the state i is said to be positive recurrent, if $m_i = \infty$ then the state i is said to be null recurrent.

Definition 1.20 Let $\{X_n; n \in \mathbb{N}\}$ be a homogeneous Markov chain and $\varnothing \neq A \subseteq S$. The arrival time to A is defined as follows:

$$T_A := \min \{n \geq 1 : X_n \in A\} \qquad (1.18)$$

if there exists any n for which $X_n \in A$. If there does not exist any n for which $X_n \in A$ then $T_A := \infty$. Note that if $A = \{i\}$ with $i \in S$ then it is written $T_A = T_i$.

Theorem 1.2 *Let $\{X_n; n \in \mathbb{N}\}$ be a homogeneous Markov chain. For all i, $j \in S$ and $n \geq 1$ it is satisfied that:*

$$p_{ij}^{(n)} = \sum_{k=1}^{n} P(T_j = k \mid X_0 = i) p_{jj}^{n-k} \qquad (1.19)$$

Definition 1.21 Let $\{X_n; n \geq 1\}$ be a homogeneous Markov chain. The probability that the chain, starting at i, will visit state j is defined as follows:

$$f_{ij} := P\left(T_j < \infty \mid X_0 = i\right) \qquad (1.20)$$

Remark 1.5 It is clear that $f_i = f_{ii}$. Also

$$f_{ij} = \sum_{k=1}^{\infty} P\left(T_j = k \mid X_0 = i\right) \qquad (1.21)$$

Definition 1.22 Let $\{X_n; n \geq 1\}$ be a homogeneous Markov chain. We define the duration of permanence of the chain in the state $j \in S$ as follows:

$$H_j := \inf \{n > 0 : X_n \neq j\} \qquad (1.22)$$

Proposition 1.3 *Let $\{X_n; n \geq 1\}$ be a homogeneous Markov chain and $i \in S$, then for all $n \in \mathbb{N}$, we have*

$$P(H_i = n \mid X_0 = i) = (p_{ii})^n (1 - p_{ii}). \tag{1.23}$$

Proof

$$
\begin{aligned}
P(H_i = n \mid X_0 = i) &= P(X_1 = i, X_2 = i, \ldots, X_{n-1} = i, X_n = i, X_{n+1} \neq i \mid X_0 = i) \\
&= P(X_1 = i \mid X_0 = i) P(X_2 = i \mid X_1 = i) \cdots P(X_{n+1} \neq i \mid X_n = i) \\
&= (p_{ii})^n (1 - p_{ii}).
\end{aligned}
$$

We see that H_i has a geometric distribution with parameter p_{ii}.

Remark 1.6 Consider a homogeneous Markov chain $\{X_n; n \geq 1\}$ and $i, j \in S$. Suppose that N_{ij} denotes the number of visits the chain makes to state j having started from i. Then for all $m \geq 1$ is satisfied:

$$P(N_{ij} > m \mid X_0 = i) = f_{ij} (f_{jj})^m \tag{1.24}$$

Theorem 1.3 *Let $\{X_n; n \geq 1\}$ be a homogeneous Markov chain and $\varrho_{ij}(m, n) :=$ "probability that the chain visits state j for the first time at time m and that the next visit to state j occurs exactly n time units later". Then:*

$$\varrho_{ij}(m, n) = P(T_j = m \mid X_0 = i) P(T_j = n \mid X_0 = j). \tag{1.25}$$

Definition 1.23 Let $\{X_n; n \geq 1\}$ be a homogeneous Markov chain. We define

$$p_{ij}^* := \sum_{n=1}^{\infty} p_{ij}^{(n)} \tag{1.26}$$

Remark 1.7 Let $\{X_n; n \geq 1\}$ be a homogeneous Markov chain. Then

$$
\begin{aligned}
\sum_{n=1}^{\infty} p_{ij}^{(n)} &= \sum_{n=1}^{\infty} P(X_n = j \mid X_0 = i) \\
&= \sum_{n=1}^{\infty} E\left(X_{\{X_n = j\}} \mid X_0 = i \right) \\
&= E\left(\sum_{n=1}^{\infty} X_{\{X_n = j\}} \mid X_0 = i \right).
\end{aligned}
$$

Therefore p_{ij}^* is the expected number of visits the chain makes to state j having started from state i.

Let $\{X_n; n \geq 1\}$ be homogeneous Markov chain. Then for all $i, j \in S$ it has to:

$$f_{ij} = \sum_{k=1}^{\infty} P\left(T_j = k \mid X_0 = i\right) \tag{1.27}$$

Then

$$p_{ij}^* = \sum_{n=1}^{\infty} \sum_{k=1}^{n} P\left(T_j = k \mid X_0 = i\right) p_{jj}^{n-k}$$

$$= \sum_{k=1}^{\infty} P\left(T_j = k \mid X_0 = i\right) \sum_{l=0}^{\infty} p_{jj}^l$$

$$= f_{ij}\left[1 + p_{jj}^*\right]$$

Therefore if $i = j$, then

$$p_{jj}^* = f_j\left[1 + p_{jj}^*\right]$$

This is,

$$p_{jj}^* = \frac{f_j}{1 + f_j}$$

$$p_{ij}^* = f_{ij}\left[1 + \frac{f_j}{1 - f_j}\right] = \frac{f_{ij}}{1 - f_j} \tag{1.28}$$

We see that

$$p_{jj}^* = \begin{cases} \infty & if \ j \quad \text{is recurrent} \\ < \infty & if \ j \quad \text{is transitory} \end{cases} \tag{1.29}$$

The following result asserts that transience and recurrence are class properties, that is, if the states are communicating, then they are of the same type.

Proposition 1.4 *Let* $\{X_n; n \in \mathbb{N}\}$ *be a homogeneous Markov chain. If* $i, j \in S$ *such that* $i \longleftrightarrow j$. *Then* i *is recurrent if and only if* j *is recurrent.*

Proof Since $i \leftrightarrow j$ exist $m, n \in \mathbb{N}$ such that $p_{ij}^n > 0$ and $p_{ji}^m > 0$. Then

$$p_{ij}^* = \sum_{k=1}^{\infty} p_{jj}^k$$

$$\geq \sum_{k=1}^{\infty} p_{jj}^{n+m+k}$$

$$\geq \sum_{k=1}^{\infty} p_{ji}^m p_{ii}^k p_{ij}^n$$

This is,

$$p_{ij}^* \geq p_{ji}^m p_{ij}^* p_{ij}^n$$

Therefore:

$$p_{ii}^* = \infty \Leftrightarrow p_{jj}^* = \infty.$$

Hence the theorem is proved.   □

From the above result, we can easily prove the following the proposition

Proposition 1.5 *Let $\{X_n; n \in \mathbb{N}\}$ be a homogeneous and irreducible Markov chain. Then one and only of the following condition is satisfied:*

1. *All states are positive recurrent.*
2. *All states are null recurrent.*
3. *All states are transitory.*

Proposition 1.6 *Let $\{X_n; n \in \mathbb{N}\}$ be a Markov chain with a finite set of states S. Then there are sets pairwise disjoint $\mathcal{T}, R_1, R_2, \ldots, R_l$ with*

$$S = \mathcal{T} \cup R_1 \cup R_2 \cup \cdots \cup R_l$$

with all the states $i \in \mathcal{T}$ are transient and the sets $R_1, R_2, \ldots, R_l$ are closed sets and irreducible.

Proof Consider $\mathcal{T} := \{i \in S : \exists j \in S \text{ with } i \longrightarrow j \wedge j \nrightarrow i\}$. It is clear that all states of $\mathcal{T}$ they are transient. Let $i_1 \in S - \mathcal{T}$. The set

$$R_1 := \{j \in S : i_1 \longrightarrow j\}$$

is closed and irreducible. We have

a. If $j \in R_1$ and $k \in S$ with $j \longrightarrow k$, $i_1 \longrightarrow j \longrightarrow k$ and therefore $k \in R_1$.
b. If $j, k \in R_1$ given that $i_1 \notin \mathcal{T}$. Then $j \longrightarrow i_1$. Consequently $j \longrightarrow k$. Analogously, it is proved that $k \longrightarrow j$.

If $S - (R_1 \cup \mathcal{T}) = \phi$, then $S = R_1 \cup \mathcal{T}$ and the result is obvious.
If $S - (R_1 \cup \mathcal{T}) \neq \phi$ then choose a $i_2 \in S - (R_1 \cup \mathcal{T})$ and it defines:

$$R_2 := \{j \in S : i_2 \longrightarrow j\}.$$

The set R_2 is irreducible and closed. Again, there are two possibilities, either $S - (R_1 \cup R_2 \cup \mathcal{T}) = \phi$ or $S - (R_1 \cup R_2 \cup \mathcal{T}) \neq \phi$. In the first case, the result of the proposition is obtained; in the second, the procedure described above is repeated. As S is finite, it is possible to construct a sequence of sets pairwise disjoint, irreducible, and closed $R_1, R_2, \ldots, R_l$ such that:

$$S = \mathcal{T} \cup R_1 \cup R_2 \cup \cdots \cup R_l.$$

Example 1.6 Consider the Markov chain $\{X_n; n \in \mathbb{N}\}$ with a set of states $S = \{a, b, c, d, e, f, g\}$ and transition matrix $\mathbf{P} = (p_{ij})_{i,j \in S}$ with probabilities are given by

$$p_{ij} = \begin{cases} 1 & \text{if } i = a, j = e; i = c, j = d; i = d, j = g; i = e, j = a; i = f, j = c; i = g, j = c \\ \frac{1}{3} & \text{if } \qquad\qquad\qquad i = b, j = a; i = b, j = c; i = b, j = f \\ 0 & \text{in} \qquad\qquad\qquad\qquad\qquad \text{otherwise} \end{cases}$$

We obtain that $S = \mathcal{T} \cup R_1 \cup R_2$ with $\mathcal{T} = \{b, f\}$, $R_1 = \{a, e\}$ and $R_2 = \{c, d, g\}$.

Definition 1.24 Let $\{X_n; n \in \mathbb{N}\}$ be a homogeneous Markov chain. The period of the state $i \in S$, $d(i)$ is defined as follows:

$$d(i) := G.C.D. \{n \geq 1 : p_{ii}^n > 0\} \tag{1.30}$$

where $G.C.D.$ denotes the greatest common divisor.

Proposition 1.7 *Let $\{X_n; n \in \mathbb{N}\}$ be a homogeneous Markov chain. If $i, j \in S$ such that $i \leftrightarrow j$, then $d(i) = d(j)$.*

Proof Since $i \leftrightarrow j$ then exist $k, m \geq 1$ such that $p_{ij}^k > 0$ and $p_{ji}^m > 0$. Let $n \geq 1$ with $p_{jj}^n > 0$. Then it is obtained that $p_{ii}^{m+k} > 0$ and $p_{ii}^{m+n+k} > 0$. Therefore $d(i)$ divides both to $(m+k)$ like $(m+n+k)$ and in consequence $d(i)$ divide to n. Then $d(i) \leq d(j)$. Similarly it is shown that $d(j) \leq d(i)$. □

The period of state i is concerned with the times at which the chain might have returned to state i. State i is called aperiodic when $d(i) = 1$. A state i is called periodic with period $k > 1$ when $d(i) = k$. If for all $n \geq 1$, $p_{ii}^n = 0$, then we define $d(i) := 0$. A homogeneous Markov chain is said to be aperiodic if all states are aperiodic; otherwise, if all states are periodic, the chain is said to be periodic. If the chain is irreducible, aperiodic, and all its states are positive recurrent, then it is said to be an ergodic chain.

Theorem 1.4 *Let* $\{X_n; n \in \mathbb{N}\}$ *be an ergodic Markov chain then*

$$\lim_{n \to \infty} p_{ij}^n = \frac{1}{m_j} \tag{1.31}$$

regardless of starting state i.

Definition 1.25 Let $\{X_n; n \geq 1\}$ be a homogeneous Markov chain. The chain is said to be absorbing if it has at least one state $i \in S$ absorbing, that is, a state i for which $p_{ii} = 1$.

Theorem 1.5 *Let* $\{X_n; n \in \mathbb{N}\}$ *a homogeneous Markov chain. Then*

1. *If* $\{X_n; n \in \mathbb{N}\}$ *is aperiodic, then the chain has a limit distribution.*
2. *If* $\{X_n; n \in \mathbb{N}\}$ *is irreducible and aperiodic, then the limit distribution is independent of the initial distribution.*
3. *If* $\{X_n; n \in \mathbb{N}\}$ *is ergodic, then the limit distribution is stationary and unique. This distribution is obtained by solving the equation*

$$\pi = \pi P \tag{1.32}$$

Furthermore, the ith component of π *is given by:*

$$\pi_i = \frac{1}{m_i} \tag{1.33}$$

where m_i *is the mean recurrence time of the state* i.

Example 1.7 Let $\{X_n; n \in \mathbb{N}\}$ be a Markov chain with set of states $S = \{0, 1\}$ and transition matrix is given by

$$P = \begin{pmatrix} \frac{2}{3} & \frac{1}{3} \\ \frac{2}{5} & \frac{3}{5} \end{pmatrix}$$

We have that $\pi = \left(\frac{6}{11}, \frac{5}{11} \right)$ for the stationary matrix P.

The stationary distribution of a Markov chain may not exist, and if it does exist, it may not be unique.

Example 1.8 Let $\{X_n; n \in \mathbb{N}\}$ be a Markov chain with states space $S = \{1, 2, 3\}$ and transition matrix P is given by:

$$P = \begin{pmatrix} 1 & 0 & 0 \\ \frac{1}{3} & \frac{1}{3} & \frac{1}{3} \\ 0 & 0 & 1 \end{pmatrix}$$

We have for each α between 0 and a 1, the vector $\pi = (1 - \alpha, 0, \alpha)$ is stationary over S.

Example 1.9 Let $\{X_n; n \in \mathbb{N}\}$ be a symmetric random walk on the integers. This chain has no stationary distribution.

Theorem 1.6 *Let $\{X_n; n \in \mathbb{N}\}$ be an irreducible aperiodic Markov chain with the state set $S \subseteq \mathbb{N}$. A probability invariant measure exists on S, if and only if the chain is positive recurrent. That probability measure is uniquely determined and the condition given in the Eq. (1.32).*

Remark 1.8 It is known that if $\{X_n; n \in \mathbb{N}\}$ is an irreducible Markov chain with finite set of states S then not all states of $\{X_n; n \in \mathbb{N}\}$ can be transient and the chain cannot have null recurrent states. Consequently, the chain is positive recurrent and therefore there must exist an stationary probability measure on S.

Example 1.10 Consider a Markov chain $\{X_n; n \in \mathbb{N}\}$ with state space $S = \{1, 2, 3\}$ and transition matrix.

$$P = \begin{pmatrix} 0 & \frac{3}{4} & \frac{1}{4} \\ \frac{1}{2} & 0 & \frac{1}{2} \\ 1 & 0 & 0 \end{pmatrix}$$

The process $\{X_n; n \in \mathbb{N}\}$ is an irreducible and aperiodic Markov chain, since also S is finite then the chain is positive recurrent and consequently there is a stationary probability measure $\pi = (\pi_j)_{j \in S}$ over S.

Since $\pi P = \pi$, then we obtain the system:

$$\begin{cases} \pi_1 + \pi_2 + \pi_3 = 1 \\ \pi_1 = \frac{1}{2}\pi_2 + \pi_3 \\ \pi_2 = \frac{3}{4}\pi_1 \\ \pi_3 = \frac{1}{4}\pi_1 + \frac{1}{2}\pi_2 \end{cases}$$

Solving, we get $\pi_1 = \frac{8}{19}$, $\pi_2 = \frac{6}{19}$, and $\pi_3 = \frac{5}{19}$.
We know that

$$\lim_{n \to \infty} p_{ij}^{(n)} = \pi_j$$

For $j = 1, 2, 3$ independent of i, this implies, in particular, that the probability that for n large enough, the chain will be at 1 given that it started from i is equal to $\frac{8}{19}$, independent of the starting state i.

Example 1.11 In this example, the daily cases of COVID-19 in Colombia from March 6, 2020, to November 30, 2022, were analyzed to the long-term probability behavior of the daily reported infection cases. Let $\{X_n; n \geq 0\}$ be a Markov chain with state space $S = \{1, 2\}$, state 1 represents the daily cases of infection increase from the previous day, and state 2 represents the daily cases of infection decrease from the previous day. We obtain the following transition matrix of COVID-19 using Python programming:

$$\mathbf{P} = \begin{pmatrix} 0.49281314 & 0.50718686 \\ 0.48425197 & 0.51574803 \end{pmatrix}$$

As we calculated the stationary probabilities earlier, we obtain (π) using $\pi = \pi\mathbf{P}$, then the stationary probabilities are given by:

$$\pi_1 = 0.48843353 \quad and \quad \pi_2 = 0.51156647.$$

From this example, we can say that the long-term reported daily infection will decrease by approximately 51.2%, and the increase is nearly 48.8%.

Example 1.12 Consider a Markov chain with state space $S = \{0, 1, 2, 3, \dots\}$, for which starting from i and makes transition one step to the states $(i - 1)$ and $(i + 1)$. If we suppose, for example, that i represents the number of individuals in a population then the transition on $i \mapsto i + 1$ represents a birth and the transition $i \mapsto i - 1$ a death. Suppose $p_{i,i-1} = \mu_i$ and $p_{i,i+1} = \lambda_i$ with $\lambda_i + \mu_i = 1, 0 < \lambda_i < 1, \mu_0 = 0, i = 0, 1, 2, \dots$. Let's assume that there is a stationary probability measure π on S. In this case we obtain the system:

$$\begin{cases} \pi_0 = \mu_1\pi_1 \\ \pi_j = \mu_{j+1}\pi_{j+1} + \lambda_{j-1}\pi_{j-1}, j \geq 1 \end{cases}$$

If we know π_0 we can determine π_j recurrently:

$$\begin{cases} \pi_1 = \frac{\pi_0}{\mu_1} \\ \pi_2 = \pi_0 \cdot \frac{(\lambda_0\lambda_1)}{(\mu_1\mu_2)} \end{cases}$$

Inductively we obtain that:

$$\pi_j = \pi_0 \prod_{k=0}^{j-1} \left(\frac{\lambda_k}{\mu_{k+1}} \right)$$

It then follows that π is a stationary probability measure on S, if and only if,

$$\sigma := \sum_{j=1}^{\infty} \prod_{k=0}^{j-1} \left(\frac{\lambda_k}{\mu_{k+1}} \right) < \infty$$

with $\pi_0 := \frac{1}{1+\sigma}$.

Corollary 1.1 If $\{X_n; n \in \mathbb{N}\}$ is an ergodic Markov chain with finite set of states $S = \{0, 1, 2, \dots, m\}$. Then there is a unique stationary distribution π and with

$$\lim_{n\to\infty} \mathbf{P}^n = \begin{pmatrix} \pi_0 & \pi_1 & \cdots & \pi_m \\ \pi_0 & \pi_1 & \cdots & \pi_m \\ \vdots & \vdots & \ddots & \vdots \\ \pi_0 & \pi_1 & \cdots & \pi_m \end{pmatrix} \tag{1.34}$$

and

$$\lim_{n\to\infty} P\left(X_n = j\right) = \pi_j \tag{1.35}$$

Proof The existence of the stationary distribution π is guaranteed by the previous theorem. On the other hand, we have:

$$\lim_{n\to\infty} P\left(X_n = j\right) = \lim_{n\to\infty} \sum_{i=0}^m P\left(X_0 = i\right) P\left(X_n = j \mid X_0 = i\right)$$

$$= \sum_{i=0}^m P\left(X_0 = i\right) \lim_{n\to\infty} P\left(X_n = j \mid X_0 = i\right)$$

$$= \pi_j \sum_{i=0}^m P\left(X_0 = i\right) = \pi_j. \qquad \square$$

1.3   Finite Markov Chain

Let $\{X_n; n \in \mathbb{N}\}$ be a homogeneous Markov chain with finite states $S = \{1, 2, \ldots, m\}$. Suppose that r of the states are recurrent and $(m - r)$ are transient. In this case the transition matrix $\mathbf{P}$ of the chain can be expressed as follows:

$$\mathbf{P} = \begin{pmatrix} \mathbf{P}_1 & 0 \\ A & Q \end{pmatrix} \tag{1.36}$$

where $\mathbf{P}_1$ denotes the transition matrix between recurring states, A is the transition matrix from transient to recurrent states, Q is the transition matrix between transient states, and 0 denotes the null matrix of size $r \times (m - r)$.

Proposition 1.8 *Let $\{X_n; n \in \mathbb{N}\}$ be a homogeneous Markov chain with finite set of states $S = \{1, 2, \ldots, m\}$ and suppose that the transition matrix is partitioned as in (1.36). Then the matrix $(I - Q)$ is invertible where I is the identity matrix of order $(m - r) \times (m - r)$.*

Proof Since:

$$I - Q^n = (I - Q)\left(I + Q + Q^2 + \cdots + Q^n\right)$$

and

$$\lim_{n \to \infty} Q^n = 0$$

Then for n large enough, we have

$$\det \left((1 - Q)\left(I + Q + Q^2 + \cdots + Q^n\right)\right) \neq 0$$

and in consequence $(I - Q)$ is invertible and we get

$$(I - Q)^{-1} = \sum_{k=0}^{\infty} Q^k.$$

Definition 1.26 Let $\{X_n; n \in \mathbb{N}\}$ be a homogeneous Markov chain with finite set of states $S = \{1, 2, \ldots, m\}$ and $\mathbf{T}$ the set of transitory states. Suppose $i, j \in \mathbf{T}$ and that the chain starts from i. If N_{ij} is the random variable denoting the number of times the chain visits state j before reaching, possibly a recurrent state, define $\mu_{ij} := E\left(N_{ij}\right)$ and $\mathbf{M} := \left(\mu_{ij}\right)_{i,j \in \mathbf{T}}$.

Theorem 1.7 *Let* $\{X_n; n \in \mathbb{N}\}$ *be a homogeneous Markov chain with a finite set of states* $S = \{1, 2, \ldots, m\}$ *and suppose that* $\mathbf{T}$ *denotes the set of transient states. Then*

$$\mathbf{M} = (I - Q)^{-1} \tag{1.37}$$

where I denotes the identity matrix of size $(m - r) \times (m - r)$.

Proof

$$\mu_{ij} = \delta_{ij} + \sum_{k \in \mathbf{T}} p_{ik} \mu_{kj}$$

where $\delta_{ij} = 1$ if $i = j$ and $\delta_{ij} = 0$ if $i \neq j$. This is,

$$\mathbf{M} = I + Q\mathbf{M}$$

$$\mathbf{M} = (I - Q)^{-1}.$$

The matrix $\mathbf{M}$ given in the previous theorem is also called the fundamental matrix.

Remark 1.9 Note that for $i \in \mathbf{T}$

$$\tau_i = \sum_{j \in \mathbf{T}} \mu_{ij} \tag{1.38}$$

denotes the average number of transitions the chain makes before leaving the transient class, having started from the transient state i.

Definition 1.27 Let $\{X_n; n \in \mathbb{N}\}$ be a homogeneous Markov chain with a finite set of states $S = \{1, 2, \ldots, m\}$. The state i is a transitory state and j a recurrent state. Define:

$$\Upsilon_{ij} := \min \{n \in \mathbb{N} : X_n = j \mid X_0 = i\} \tag{1.39}$$

that is, Υ_{ij} denotes the minimum number of transitions that the chain requires, having left the transient class, to reach the recurrent state j.

Let $(X_n)_{n \in \mathbb{N}}$ be a homogeneous discrete-time Markov chain with a finite set of states $S = \{1, 2, \ldots, m\}$. We define for $i \in \mathbf{T}$, and $j \in \mathbf{T}^c$

$$g_{ij}^n := P\left(\Upsilon_{ij} = n\right)$$

$$g_{ij} := \sum_{n=1}^{\infty} g_{ij}^n$$

Write $G^n := \left(g_{ij}^n\right)_{i \in \mathbf{T}, j \in \mathbf{T}^c}$ and $G = \left(g_{ij}\right)_{i \in \mathbf{T}, j \in \mathbf{T}^c}$.

Theorem 1.8 *Let $\{X_n; n \in \mathbb{N}\}$ be a homogeneous Markov chain with a finite set of states $S = \{1, 2, \ldots, m\}$. Then:*

$$G^n = Q^{n-1} A \; G = \mathbf{M} A. \tag{1.40}$$

Proof It is clear that $G^1 = A$ and $G^n = QG^{n-1}$. Then for $n \geq 2$;

$$G^n = QG^{n-1} = QQG^{n-2} = \cdots = Q^{n-1} A$$

On the other hand,

$$G = \sum_{n=1}^{\infty} G^n$$

$$= A + \sum_{n=2}^{\infty} Q^{n-1} A$$

$$= A + QA + Q^2 A + Q^3 A + \cdots$$

$$= \left(I + Q + Q^2 + \cdots\right) A = \mathbf{M} A. \qquad \square$$

Corollary 1.2 *Let $X = \{X_n; n \in \mathbb{N}\}$ be a homogeneous Markov chain with a finite set of states $S = \{1, 2, \ldots, m\}$. Then the probabilities that, starting from any transitory state i, the chain reaches the recurrent state j are given by $\mathbf{M} A_j$, where A_j denotes the jth column of the matrix A.*

Example 1.13 A player comes to a casino with 200. He decides to play until his capital equals 700 or until he goes broke. In each round of the game, the player wins or loses 100 with probabilities 0.4 and 0.6 respectively. What is the probability that the player reaches his goal of 700? What is the probability of player ruin?

Solution: Let X_n be the random variable denoting the player's fortune after the nth game. According to the data of the problem, we have that the set of states of the chain $\{X_n; n \geq 0\}$ with $S = \{0, 1, \ldots, 7\}$, and where $k \in S$ denotes the player's fortune in hundreds of dollars. The transition matrix of the chain is equal to

$$P = \begin{pmatrix} 1 & 0 & 0 & 0 & 0 & 0 & 0 & 0 \\ 0.6 & 0 & 0.4 & 0 & 0 & 0 & 0 & 0 \\ 0 & 0.6 & 0 & 0.4 & 0 & 0 & 0 & 0 \\ 0 & 0 & 0.6 & 0 & 0.4 & 0 & 0 & 0 \\ 0 & 0 & 0 & 0.6 & 0 & 0.4 & 0 & 0 \\ 0 & 0 & 0 & 0 & 0.6 & 0 & 0.4 & 0 \\ 0 & 0 & 0 & 0 & 0 & 0.6 & 0 & 0.4 \\ 0 & 0 & 0 & 0 & 0 & 0 & 0 & 1 \end{pmatrix}$$

This chain has 3 equivalence classes:

$$C(0) = \{0\}$$
$$C(1) = \{1, 2, \ldots, 6\}$$
$$C(7) = \{7\}$$

The states 0 and 7 are recurrent (also absorbing) states and other states are transient states. We want to calculate G_{27}. According to the theory developed, this probability corresponds to the component located in the second row and second column of matrix $G = MA$ where

$$M = (I - Q)^{-1}$$

$$= \left(\begin{pmatrix} 1 & 0 & 0 & 0 & 0 & 0 \\ 0 & 1 & 0 & 0 & 0 & 0 \\ 0 & 0 & 1 & 0 & 0 & 0 \\ 0 & 0 & 0 & 1 & 0 & 0 \\ 0 & 0 & 0 & 0 & 1 & 0 \\ 0 & 0 & 0 & 0 & 0 & 1 \end{pmatrix} - \begin{pmatrix} 0 & 0.4 & 0 & 0 & 0 & 0 \\ 0.6 & 0 & 0.4 & 0 & 0 & 0 \\ 0 & 0.6 & 0 & 0.4 & 0 & 0 \\ 0 & 0 & 0.6 & 0 & 0.4 & 0 \\ 0 & 0 & 0 & 0.6 & 0 & 0.4 \\ 0 & 0 & 0 & 0 & 0.6 & 0 \end{pmatrix} \right)^{-1}$$

$$= \begin{pmatrix} 1.6149 & 1.0248 & 0.63137 & 0.36911 & 0.19427 & 7.7708 \times 10^{-2} \\ 1.5372 & 2.5619 & 1.5784 & 0.92278 & 0.48567 & 0.19427 \\ 1.4206 & 2.3677 & 2.999 & 1.7533 & 0.92278 & 0.36911 \\ 1.2458 & 2.0763 & 2.6299 & 2.999 & 1.5784 & 0.63137 \\ 0.98349 & 1.6391 & 2.0763 & 2.3677 & 2.5619 & 1.0248 \\ 0.59009 & 0.98349 & 1.2458 & 1.4206 & 1.5372 & 1.6149 \end{pmatrix}$$

and

$$
A = \begin{pmatrix} 0.6 & 0 \\ 0 & 0 \\ 0 & 0 \\ 0 & 0 \\ 0 & 0 \\ 0 & 0.4 \end{pmatrix}.
$$

We get

$$
G = \begin{pmatrix} 1.6149 & 1.0248 & 0.63137 & 0.36911 & 0.19427 & 7.7708 \times 10^{-2} \\ 1.5372 & 2.5619 & 1.5784 & 0.92278 & 0.48567 & 0.19427 \\ 1.4206 & 2.3677 & 2.999 & 1.7533 & 0.92278 & 0.36911 \\ 1.2458 & 2.0763 & 2.6299 & 2.999 & 1.5784 & 0.63137 \\ 0.98349 & 1.6391 & 2.0763 & 2.3677 & 2.5619 & 1.0248 \\ 0.59009 & 0.98349 & 1.2458 & 1.4206 & 1.5372 & 1.6149 \end{pmatrix} \begin{pmatrix} 0.6 & 0 \\ 0 & 0 \\ 0 & 0 \\ 0 & 0 \\ 0 & 0 \\ 0 & 0.4 \end{pmatrix}
$$

$$
= \begin{pmatrix} 0.96894 & 3.1083 \times 10^{-2} \\ 0.92232 & 7.7708 \times 10^{-2} \\ 0.85236 & 0.14764 \\ 0.74748 & 0.25255 \\ 0.59009 & 0.40992 \\ 0.35405 & 0.64596 \end{pmatrix}
$$

The required probability is equal to 7.7708×10^{-2}.
The player's ruin probability is given by

$$
(0, 1, 0, 0, 0, 0) \begin{pmatrix} 1.6149 & 1.0248 & 0.63137 & 0.36911 & 0.19427 & 7.7708 \times 10^{-2} \\ 1.5372 & 2.5619 & 1.5784 & 0.92278 & 0.48567 & 0.19427 \\ 1.4206 & 2.3677 & 2.999 & 1.7533 & 0.92278 & 0.36911 \\ 1.2458 & 2.0763 & 2.6299 & 2.999 & 1.5784 & 0.63137 \\ 0.98349 & 1.6391 & 2.0763 & 2.3677 & 2.5619 & 1.0248 \\ 0.59009 & 0.98349 & 1.2458 & 1.4206 & 1.5372 & 1.6149 \end{pmatrix} \begin{pmatrix} 0.6 \\ 0 \\ 0 \\ 0 \\ 0 \\ 0 \end{pmatrix}
$$

$$
= 0.92232.
$$

Maximum-Likelihood Estimation for Markov Chains

We now briefly explain the maximum-likelihood method to estimate transition probabilities. Suppose that $x = \{x_1, x_2, \ldots, x_n\}$ is a realization of length n, of a homogeneous Markov chain $\{X_n; n \in \mathbb{N}\}$ with finite set of states $S = \{1, 2, \ldots, m\}$. In this case, the likelihood function is given by:

$$L\left(p_{ij}; x\right) \propto P\left(X = x; p_{ij}\right)$$

$$= P\left(X_0 = i_0\right) \prod_{k=1}^{n} p_{x_{k-1}x_k}$$

$$= P\left(X_0 = i_0\right) \prod_{i,j=1}^{m} p_{ij}^{n_{ij}}$$

where n_{ij} is the number of transitions from state i to state j. Then

$$\log\left(L\left(p_{ij}; x\right)\right) = \log\left(P\left(X_0 = i_0\right)\right) + \sum_{i,j=1}^{m} n_{ij} p_{ij} \qquad (1.41)$$

Subject to restriction

$$\sum_{j=1}^{m} p_{ij} = 1 \qquad (1.42)$$

and using Lagrange multipliers we get

$$l\left(p_{ij}, x\right) = \log\left(P\left(X_0 = i_0\right)\right) + \sum_{i,j=1}^{m} n_{ij} \log p_{ij} + \lambda\left(1 - \sum_{j=1}^{m} p_{ij}\right)$$

Differentiating l with respect to p_{ij} and λ gives

$$\frac{\partial l}{\partial p_{ij}} = \frac{n_{ij}}{p_{ij}} - \lambda$$

and

$$\frac{\partial l}{\partial \lambda} = 1 - \sum_{j=1}^{m} p_{ij}$$

Then, equating both expressions to zero, we obtain:

$$\lambda = \frac{n_{ij}}{p_{ij}}$$

and

$$1 = \sum_{j=1}^{m} p_{ij} = \frac{1}{\lambda} \sum_{j=1}^{m} n_{ij}$$

This is,

$$p_{ij} = \frac{n_{ij}}{\sum_{j=1}^{m} n_{ij}}$$

Therefore the maximum likelihood estimator of the transition probability p_{ij} is given by:

$$\widehat{p_{ij}} = \frac{n_{ij}}{n_i}$$

where

$$n_i = \sum_{j=1}^{m} n_{ij}.$$

Example 1.14 Paddy rice cultivation in Colombia is important to the country's economy, according to information from the Ministry of Agriculture and Rural Development(Minagricultura) of Colombia. Rice is grown in approximately 20% of the country's municipalities, in some of them, this crop represents up to 80% of the family income. In Colombia, rice cultivation generates 75,000 direct jobs and 300,000 indirect jobs. Different authors have used the Markov chain theory to model the fluctuation of cereal prices, particularly rice, in the market [7]. Knowing the dynamics of the cost of Paddy rice in Colombia is very useful for the rice sector because it allows them to make decisions regarding planning the planting of crops and determine the best investment strategies. In his final Master's project, our student Cristian Pérez uses data from the National Federation of Rice Growers of Colombia (Fedearroz) to build a Markov chain to determine the behavior of the monthly price of Paddy rice in the period between January 1996 and July 2022.

Let $X_n :=$ Paddy rice price in Colombia in the nth month with states space $S = \{A, B\}$, where A represents an increase in price and B represents a decrease in the cost of the rice. We the actual data, we calculate the following,

$$n_{AA} = 149, n_{AB} = 41, n_{BA} = 39, n_{BB} = 89$$

where

$$n_{ij} := \text{"number of transitions of state } i \text{ to the state } j\text{"}, i, j \in S$$

The estimate of the transition matrix is given by:

$$\widehat{P} = \begin{pmatrix} \frac{149}{190} & \frac{41}{190} \\ \frac{39}{128} & \frac{89}{128} \end{pmatrix}$$

The limit distribution of the chain π is obtained by solving the equation

$$\pi = \pi \widehat{P} \tag{1.43}$$

That is:

$$(\pi_A, \pi_B) = (\pi_A, \pi_B) \begin{pmatrix} \frac{149}{190} & \frac{41}{190} \\ \frac{39}{128} & \frac{89}{128} \end{pmatrix}$$

Solving the system of equations,

$$\frac{41}{190} \pi_A - \frac{39}{128} \pi_B = 0$$

$$-\frac{41}{190}\pi_A + \frac{39}{128}\pi_B = 0$$
$$\pi_A + \pi_B = 1$$

we get,

$$\pi_A = 0.5854$$
$$\pi_B = 0.41460$$

Therefore, the probability that the price of Paddy rice will increase is equal to 0.5854 and that it will decrease is 0.41460. The mean recurrence times of states A and B are, respectively:

$$m_A = \frac{1}{0.5854} = 1.7082$$

and

$$m_B = \frac{1}{0.41460} = 2.4120$$

We conclude that for approximately 1.7 months the price of rice remains rising and for 2.142 months the price remains decreasing.

Utility Functions

Let $(X_n)_{n\geq0}$ be a Markov chain with state set S. Let G be a function defined on S and with values in $\mathbb{R}$. The stochastic process $(Y_n)_{n\geq0}$ is defined as

$$Y_n := G(X_n).$$

We observe that Y_n is not necessarily a Markov chain as shown in the following example.

Example 1.15 Consider $(X_n)_{n\geq0}$ be a Markov chain with set of states $S = \{a, b, c\}$, transition matrix P and initial distribution $\pi = \left(\frac{1}{3}, \frac{1}{3}, \frac{1}{3}\right)$

$$P = \begin{pmatrix} \frac{1}{2} & \frac{1}{2} & 0 \\ \frac{1}{3} & 0 & \frac{2}{3} \\ \frac{1}{3} & \frac{1}{3} & \frac{1}{3} \end{pmatrix}$$

Let G be the function given by $G(a) = G(b) = 1$ and $G(c) = -1$

We have

$$P(Y_2 = 1 \mid Y_1 = 1) = \frac{P(Y_2 = 1, Y_1 = 1)}{P(Y_1 = 1)}$$

$$= \frac{1}{P\left(X_1 = a\right) + P\left(X_1 = b\right)}\left(P\left(X_2 = a, X_1 = a\right)\right.$$
$$+ P\left(X_2 = b, X_1 = a\right)$$
$$+ P\left(X_2 = a, X_1 = b\right)$$
$$\left. + P\left(X_2 = b, X_1 = b\right)\right)$$
$$= \frac{1}{P\left(X_1 = a\right) + P\left(X_1 = b\right)}\left(P\left(X_2 = a \mid X_1 = a\right) P\left(X_1 = a\right)\right.$$
$$+ P\left(X_2 = b \mid X_1 = a\right) P\left(X_1 = a\right)$$
$$+ P\left(X_2 = a \mid X_1 = b\right) P\left(X_1 = b\right)$$
$$\left. + P\left(X_2 = b \mid X_1 = b\right) P\left(X_1 = b\right)\right)$$
$$= \frac{\frac{1}{2} \times \frac{7}{18} + \frac{1}{2} \times \frac{7}{18} + \frac{1}{3} \times \frac{5}{18}}{\frac{7}{18} + \frac{5}{18}}$$
$$= 0.722\,22$$

and similarly, we get

$$P\left(Y_2 = 1 \mid Y_1 = 1, Y_0 = -1\right) = \frac{P\left(Y_2 = 1, Y_1 = 1, Y_0 = -1\right)}{P\left(Y_1 = 1, Y_0 = -1\right)}$$
$$= \frac{2}{3}.$$

Remark 1.10 Suppose that $(X_n)_{n \geq 0}$ is a discrete-time Markov chain with set of states $S = \{1, 2, \ldots, m\}$ and the transition matrix $P = \left(p_{ij}\right)_{i,j \in S}$. We have

$$E\left(Y_n \mid X_0 = i\right) = E\left(G\left(X_n\right) \mid X_0 = i\right)$$
$$= \sum_{j=1} G\left(j\right) p_{ij}^{(n)}.$$

Here, we see that $E\left(Y_n \mid X_0 = i\right)$ is the ith component of the matrix **GP** with **G** = $(G\left(1\right), G\left(2\right), \ldots, G\left(m\right))$.

Proposition 1.9 *If $(X_n)_{n \geq 0}$ is a Markov chain with set of states $S = \{1, 2, \ldots, m\}$, irreducible and aperiodic, then*

$$\lim_{n \to \infty} \frac{1}{n} \sum_{k=0}^{n} G\left(X_k\right) = \pi G^T$$

where π is the stationary distribution of the chain.

Example 1.16 Suppose that a machine on the nth day, in three possible states:

$$A := \text{"service state"}$$

$$B := \text{"failure state"}$$
$$C := \text{"repair state"}$$

Let
$$X_n := \text{"state of the machine on the } n\text{th day"}$$

Suppose that $(X_n)_{n \geq 0}$ is a Markov chain with transition matrix given by.

$$P = \begin{pmatrix} 0.9 & 0.1 & 0 \\ 0.7 & 0 & 0.3 \\ 0.5 & 0 & 0.5 \end{pmatrix}$$

and that the daily profit (utility) generated by the machine (in thousands of dollars) is given by:

$$G(A) = 1$$
$$G(B) = -1$$
$$G(C) = -3$$

The machine is said to be profitable if, in the long run, the average value of the utility is positive. Determine if the machine is profitable.

Solution: The stationary distribution π of the chain must be calculated. We have from the P matrix

$$0.1\pi_A - 0.7\pi_B - 0.5\pi_C = 0$$
$$\pi_B - 0.1\pi_A = 0$$
$$0.3\pi_B - 0.5\pi_C = 0$$
$$\pi_A + \pi_B + \pi_C = 1$$

solving the system, we get

$$\pi_A = \frac{50}{58}$$
$$\pi_B = \frac{5}{58}$$
$$\pi_C = \frac{3}{58}$$

Thus,

$$(1, -1, -3) \begin{pmatrix} \frac{50}{58} \\ \frac{5}{58} \\ \frac{3}{58} \end{pmatrix} = \frac{50}{58} - \frac{5}{58} - \frac{9}{58} = \frac{36}{58}$$

that is, the machine is profitable.

1.4   Reversible Markov Chains

The reversible Markov chains are beneficial for the construction of dynamic simulation algorithms of Markov chains. Intuitively, the reversibility of a Markov chain indicates that its evolution in time, concerning its finite-dimensional distributions, does not make any distinction between considering the behavior of the chain forwards or considering it backward. Formally we have the following definition:

Definition 1.28 Let $\{X_n; n \in \mathbb{N}\}$ be a homogeneous Markov chain with set of states S; transition matrix $\mathbf{P}$ and stationary initial distribution π_0. The chain is said to be reversible if and only if the following condition is satisfied:

$$P\left(X_0 = i_0, X_1 = i_1, \ldots, X_n = i_n\right) = P\left(X_0 = i_n, \ldots, X_n = i_0\right)$$

for all n and for all $i_0, i_1, \ldots, i_n \in S$.

Theorem 1.9 *Let $\{X_n; n \in \mathbb{N}\}$ be a homogeneous Markov chain with set of states S; transition matrix $\mathbf{P}$ and stationary initial distribution π with $\pi_i > 0$ for all $i \in S$. The chain is reversible, if and only if, for all $i, j \in S$ the so-called balance conditions are satisfied:*

$$\pi_i p_{ij} = \pi_j p_{ji}$$

Proof Since the chain is reversible then for all $i, j \in S$ it has to:

$$P\left(X_0 = i, X_1 = j\right) = P\left(X_0 = j, X_1 = i\right)$$

This is,

$$P\left(X_0 = i\right) P\left(X_1 = j \mid X_0 = i\right) = P\left(X_0 = j\right) P\left(X_1 = i \mid X_0 = j\right)$$
$$\pi_i p_{ij} = \pi_j p_{ji}$$

for all $i, j \in S$.
Now, let $n \in \mathbb{N}$ and $i_0, i_1, \ldots, i_n \in S$, then

$$
\begin{aligned}
P\left(X_0 = i_0, X_1 = i_1, \ldots, X_n = i_n\right) &= \pi_{i_0} p_{i_0 i_1} p_{i_1 i_2} \cdots p_{i_{n-1} i_n} \\
&= \pi_{i_1} p_{i_1 i_0} p_{i_1 i_2} \cdots p_{i_{n-1} i_n} \\
&= p_{i_1 i_0} \pi_{i_2} p_{i_2 i_1} \cdots p_{i_{n-1} i_n} \\
&\;\;\vdots \\
&= \pi_{i_n} p_{i_n i_{n-1}} \cdots p_{i_2 i_1} p_{i_1 i_0} \\
&= P\left(X_0 = i_n, \ldots, X_n = i_0\right).
\end{aligned}
$$

Remark 1.11 Let $\{X_n; n \geq 1\}$ be a homogeneous Markov chain with set of states S, the transition matrix $\mathbf{P}$ and stationary initial distribution π with $\pi_i > 0$ for all $i \in S$. Since $\mathbf{P}$ is a stochastic matrix then

$$\pi_i = \pi_i \sum_{j \in S} p_{ij} = \sum_{j \in S} \pi_i p_{ij} = \sum_{j \in S} \pi_j p_{ji}$$

That is,

$$\pi = \pi \mathbf{P}.$$

Consequently, any initial distribution π that satisfies the balance conditions is a stationary distribution.

Theorem 1.10 (Strong law of large numbers for Markov chains)

Let $\{X_n; n \geq 1\}$ be an irreducible Markov chain with stationary initial distribution π and set of states S. The function $f : S \longrightarrow \mathbb{R}$ with

$$\sum_{j \in S} |f(j)| \pi_j < \infty$$

Then

$$\frac{1}{n} \sum_{k=1}^{n} f(X_k) \xrightarrow[n \to \infty]{c.s} \sum_{j \in S} f(j) \pi_j.$$

Corollary 1.3 *Let $\{X_n; n \geq 1\}$ be an irreducible Markov chain with stationary initial distribution π and set of states S. Then for all $i \in S$ We have*

$$\frac{N_n(i)}{n} \xrightarrow[n \to \infty]{c.s} \pi_i$$

where $N_n(i) := $ *"number of visits the chain makes to state i up to time n."*

Proof Let $i \in S$ fixed and f the function defined on S by:

$$f(k) := \begin{cases} 1 \text{ if } i = k \\ 0 \text{ if } i \neq k \end{cases}$$

Then

$$\sum_{k=1}^{n} f(X_k) = N_n(i)$$

and

$$\frac{1}{n} \sum_{k=1}^{n} f(X_k) = \frac{N_n(i)}{n}$$

corresponds to the relative frequency with which the chain visits the state i. $\qquad \square$

An irreducible, reversible finite Markov chain is ergodic. However exists examples of reversible Markov that are not ergodic chain, this as shown in the following example.

Example 1.17 (*Ehrenfest process*) Suppose that there are l particles distributed in two containers A and B connected to each other but isolated from the outside world. At time $t = n - 1$ there are i particles in the container A. We select randomly one of the particles from a container A with probability $\frac{i}{l}$, and placed in the container B. If the selected particle is in container B, with probability $\frac{l-i}{l}$, then selected particle is placed in urn A.

The $X_n :=$ "number of particles in container A at time n". The process $(X_n)_{n \geq 0}$ is a Markov chain with set of states $S = \{0, 1, 2, \ldots, l\}$ and transition matrix $\mathbf{P} = (p_{ij})_{l \times l}$ with

$$
p_{ij} = \begin{cases} \frac{l-i}{l} & \text{if } i < l \text{ and } j = i+1 \\ \frac{i}{l} & \text{if } i > 0 \text{ and } j = i-1 \\ 0 \text{ in} & \text{other case} \end{cases}
$$

The finite chain is a periodic with period $d = 2$. We easily check that even though the chain is not ergodic, it has initial stationary distribution. The equation $\pi = \pi\mathbf{P}$, has the solution $\pi = (\pi_i)_{i \in S}$ where, for $i = 0, 1, 2, \ldots, l$,

$$
\pi_i = \frac{1}{2^l} \binom{l}{i}.
$$

Since, for any $i, j \in S$, it is satisfied that

$$
\pi_i p_{ij} = \pi_j p_{ji},
$$

we see that the above chain is reversible.

Example 1.18 (*Birth and death process*) Let $\{X_n; n \in n \geq 0\}$ be a Markov chain with set of states $S = \{0, 1, 2, \ldots, l\}$ and transition matrix

$$
\mathbf{P} = \begin{pmatrix}
0 & 1 & 0 & 0 & 0 & 0 & 0 & 0 \\
q_1 & r_1 & p_1 & 0 & 0 & 0 & 0 & 0 \\
0 & q_2 & r_2 & p_2 & 0 & 0 & 0 & 0 \\
0 & 0 & \ddots & \ddots & \ddots & 0 & 0 & 0 \\
0 & 0 & 0 & q_i & r_i & p_i & 0 & 0 \\
0 & 0 & 0 & 0 & \ddots & \ddots & \ddots & 0 \\
0 & 0 & 0 & 0 & 0 & q_{l-1} & r_{l-1} & p_{l-1} \\
0 & 0 & 0 & 0 & 0 & 0 & 1 & 0
\end{pmatrix}
$$

where $p_i > 0, q_i > 0, r_i > 0$ and $p_i + q_i + r_i = 1$ for all $i = 1, 2, \ldots, l-1$.

The chain is irreducible and aperiodic. By solving the system $\pi = \pi\mathbf{P}$, that is, by solving the system of equations

$$\pi_i = \begin{cases} p_{i-1}\pi_{i-1} + r_i\pi_i + q_{i+1}\pi_{i+1} & \text{if } 0 < i < l \\ q_1\pi_1 & \text{if } i = 0 \\ p_{l-1}\pi_{l-1} & \text{if } i = l \end{cases}$$

it is obtained that:

$$\pi_0 = \left(1 + \frac{1}{q_1} + \frac{p_1}{q_1 q_2} + \cdots + \frac{p_1 \cdots p_{l-1}}{q_1 \cdots q_l}\right)^{-1}$$

and for $i = 1, 2, \ldots, l$

$$\pi_i = \pi_0 \frac{p_1 p_2 \cdots p_{i-1}}{q_1 q_2 \cdots q_i}$$

Since, for any $i, j \in S$ it is satisfied that

$$\pi_i p_{ij} = \pi_j p_{ji}$$

Then $\{X_n; n \geq 0\}$ is reversible.

In the following example, we show that there is a non-reversible ergodic Markov chain.

Example 1.19 Consider $\{X_n; n \geq 0\}$ a Markov chain with set of states $S = \{a, b, c\}$ and transition matrix given by:

$$\mathbf{P} = \begin{pmatrix} \frac{2}{5} & \frac{1}{5} & \frac{2}{5} \\ \frac{3}{5} & \frac{1}{5} & \frac{1}{5} \\ 0 & \frac{3}{5} & \frac{2}{5} \end{pmatrix}.$$

By solving the system $\pi = \pi\mathbf{P}$ is obtained $\pi = \left(\frac{1}{3}, \frac{1}{3}, \frac{1}{3}\right)$. The chain is not reversible because, for example,

$$\pi_1 p_{12} = \frac{1}{3} \times \frac{1}{5} \neq \frac{1}{3} \times \frac{3}{5} = \pi_2 p_{21}$$

Example 1.20 (*Monte Carlo method*) If X is a random variable and $g : \mathbb{R} \longrightarrow \mathbb{R}$ is a measurable function. Suppose we want to calculate (if it exists) the expected value of $Y := g(X)$. A popular method to calculate, if an approximate way, is given by the following algorithm:

Generate a sequence $\{X_n; n \geq 1\}$ of independent and equally distributed random variables with the same distribution of the random variable X.

Approximate $E(g(X))$ by the limit, when $N \to \infty$, of the arithmetic mean

$$g(\overline{X}_N) := \frac{1}{N} \sum_{i=1}^{N} g(X_i)$$

The estimator $g\left(\overline{X}_N\right)$ of $E\left(g\left(X\right)\right)$ is an unbiased estimator. However, since:

$$Var\left(g\left(\overline{X}_N\right) - E\left(g\left(X\right)\right)\right) = Var\left(g\left(\overline{X}_N\right)\right)$$

$$= \frac{1}{N^2}\sum_{i=1}^{N}Var\left(g\left(X_i\right)\right)$$

$$= \frac{1}{N}Var\left(g\left(X_1\right)\right).$$

Then the error of the approximation is of the order $O\left(\frac{1}{\sqrt{N}}\right)$. Therefore, increasing the precision of the estimate by one digit, this is, reducing its given deviation by a factor of 0.1 requires increasing the number of iterations by a factor of 100.

One way to get around this problem is to make use of the strong law of large numbers for Markov chains. The idea is then to generate a Markov chain $\{X_n; n \geq 1\}$ reversible, irreducible, and aperiodic with limiting distribution π and use

$$\frac{1}{N}\sum_{k=1}^{N}g\left(X_k\right)$$

as an estimator of $E\left(g\left(X\right)\right)$.

Then we are looking to build a Markov chain $(X_n)_{n\geq 1}$ with stationary probabilities of the form

$$\pi_i := \frac{b_i}{\sum_{j=1}^{N}b_j}$$

where the b_i are positive numbers with $i \in S := \{1, 2, \ldots, N\}$. The so-called Hasting-Metropolis algorithm offers a methodology to perform this task. The objective of Bayesian statistics is the use of information known previously to make inferences, that is, to know the properties or characteristics of unknown parameters with date previous knowledge.

Let $X = (X_1, \ldots, X_n)$ be a random vector, whose realization is an observed data set $(x_1, x_2, \ldots, x_n)$. To build a stochastic model that explains the observations obtained, it is assumed that each observation x_i comes from an unknown distribution $f\left(X_i, \theta\right)$. Under the principle that all the information regarding the parameter θ is found in the data, and the unknown value of θ is estimated using some statistics constructed from the observations. Therefore, it is assumed that the observed data corresponds to the realization of a probability distribution (called distribution a priori) $G\left(\theta\right)$. Making use of the Bayes rule determines, from the observations, the distribution of a posteriori once the data has been observed.

Rarely is it possible to find the distribution a exact posterior π of θ. Most likely, the computation of distribution a posteriori π requires difficult calculations to perform. For example, you may need to find the value of an integral that does not always have an analytic or known solution. In order to compute or approximate the distribution a posteriori π of θ, it is necessary to use methods of numerical approximation of which, a large number of variables, are not always efficient. An alternative to knowing the distribution of a posterior is through samples when its explicit form is unknown. This is possible using the Markov Chain Monte Carlo (MCMC) method. The Central idea of MCMC methods is to seek to obtain random numbers that are distributed (at least approximately) according to a certain distribution π by simulating a Markov chain whose unique stationary distribution is precise π [8].

One of the most used and well-known MCMC methods is the Metropolis-Hastings algorithm, which was initially developed by Metropolis [9]. In the year 1970, Hastings [10] extended it to the more general case. The Metropolis-Hastings algorithm seeks to simulate a Markov chain that converges to the posterior distribution π. Let S be a set of states on which is defined the target distribution π.

Suppose that

$$\overline{\Pi} = (\overline{\pi}(i, j))_{i,j \in S}$$

is a transition probability matrix, such that for each $i \in S$ it is easy, in computational terms, to generate a random sample of the distribution $\{\overline{\pi}(i, j), j \in S\}$. Then a Markov chain $\{X_n; n \in \mathbb{N}\}$ will be generated as follows: if $X_n = i$ then a sample is drawn from the distribution $\{\overline{\pi}(i, j), j \in S\}$ and is denoted as Y_n. Then, select X_{n+1} from the values X_n and Y_n, so that

$$P(X_{n+1} = Y_n \mid X_n, Y_n) = \alpha(X_n, Y_n)$$
$$P(X_{n+1} = X_n \mid X_n, Y_n) = 1 - \alpha(X_n, Y_n)$$

where $\alpha(X_n, Y_n)$ is the probability of acceptance for the sample defined by

$$\alpha(X_n, Y_n) = \alpha(i, j) := \min\left(\frac{\pi_j \overline{\pi}(j, i)}{\pi_i \overline{\pi}(i, j)}, 1\right)$$

for all $i, j \in S$ with $\pi_i \overline{\pi}(i, j) > 0$.

We have that $\{X_n; n \in \mathbb{N}\}$ is a Markov chain with probability transition matrix $(\pi(i, j))_{i,j \in S}$ where

$$\pi(i, j) := \begin{cases} \overline{\pi}(i, j) \alpha(i, j) \text{ if } i \neq j \\ 1 - \sum_{k \neq i} \pi(i, k) \text{ if } i = j \end{cases}$$

The chain $\{X_n; n \in \mathbb{N}\}$ turns out to be reversible with a stationary probability distribution π. We conclude this chapter with Metropolis-Hastings Algorithm [10].

Let π be given probability distribution. Let $q(x, y)$ be a given transition matrix. Suppose that $X_0 = \widetilde{x}$ for some value $\widetilde{x}$ with $\pi(\widetilde{x}) > 0$.

For $k = 0$ a $N - 1$ do

1. Choose a random number Y according to the probability distribution $q(X_k, .)$ and choose a random number distributed with $U \overset{d}{=} \mathcal{U}(0, 1)$.

2. Calculate

$$\alpha\left(X_k, Y\right) := \min\left(1, \frac{\pi\left(Y\right) q\left(Y, X_k\right)}{\pi\left(X_k\right) q\left(X_k, Y\right)}\right)$$

3. If

$$\alpha\left(X_k, Y\right) > U$$

then

$$X_{k+1} = Y, k = k + 1, \text{ go the step 1.}$$

Otherwise, go the step 1.

A simple example of an application of the Monte Carlo method for estimating the probability of occurrence $P\left(A\right)$ of an event A. Since

$$P\left(A\right) = E\left(1_A\right)$$

where 1_A is the random variable defined by

$$1_A\left(\omega\right) := \begin{cases} 1 \text{ if } \omega \in A \\ 0 \text{ if } \omega \notin A \end{cases}$$

then the Monte Carlo estimate of $P\left(A\right)$ consists of determining the relative frequency of occurrence of the event A in N independent randomized experiments. More precisely, if A_i denotes the occurrence of event A at the ith experiment then the relative frequency $rf\left(A\right)$ of A is equal to:

$$rf\left(A\right) = \frac{1}{N}\sum_{i=1}^{N} 1_{A_i}$$

Since

$$Var\left(A\right) = P\left(A\right) P\left(A^c\right) = P\left(A\right)\left[1 - P\left(A\right)\right]$$

then the estimation σ_N^2 the variance of A is

$$\sigma_N^2 = rf\left(A\right)\left[1 - rf\left(A\right)\right]$$

and hence, with a significance level of 95%, we obtain that $P\left(A\right)$ is in the confidence interval given by

$$\left[rf\left(A\right) - \frac{1.96}{\sqrt{N}}\sigma_N, rf\left(A\right) + \frac{1.96}{\sqrt{N}}\sigma_N\right].$$

References

1. Markov, A. A. (1906). Extension of the law of large numbers to dependent quantities. *Izv. Fiz.-Matem. Obsch. Kazan Univ.(2nd Ser)*, *15*(1), 135–156.
2. Vulpiani, A. (2015). Andrey Andreyevich Markov: A furious mathematician and his chains. *Lettera Matematica, 3,* 205–211.
3. Khmelev, D. V., & Tweedie, F. J. (2001). Using Markov chains for identification of writer. *Literary and Linguistic Computing, 16*(3), 299–307.
4. Langville, A. N., & Meyer, C. D. (2006). Updating Markov chains with an eye on Google's PageRank. *SIAM Journal on Matrix Analysis and Applications, 27*(4), 968–987.
5. Castañeda, L. B., Arunachalam, V., & Dharmaraja, S. (2012). *Introduction to probability and stochastic processes with applications* (1st ed.). New Jersey: Wiley.
6. Resnick, S. I. (2001). *Adventures in stochastic processes* (2nd ed.). Boston: Birkhäuser.
7. Respatiwulan, Prabandari, D., Susanti, Y., Handayani, S. S., & Hartatik. (2019). The stochastic model of rice price fluctuation in Indonesia. *Journal of Physics: Conference Series, 1217*(1), 012107. https://doi.org/10.1088/1742-6596/1217/1/012107.
8. Korn, R., Korn, E., & Kroisandt, G. (2010). *Monte Carlo methods and models in finance and insurance*. CRC Press.
9. Metropolis, N., Rosenbluth, A. W., Rosenbluth, M. N., Teller, A. H., & Teller, E. (1953). Equation of state calculations by fast computing machines. *The Journal of Chemical Physics, 21*(6), 1087–1092.
10. Hastings, W. K. (1970). Monte Carlo sampling methods using Markov chains and their applications.

Poisson Processes and Its Extensions

2

The Poisson process owes its name to the French physicist and mathematician Siméon Denis Poisson (1781–1840) even though he did not formulate it. The first study of Poisson processes was carried out by the English geologist John Michell (1724–1793) who was interested in determining the probability that a star was in the same region as another star, assuming that the stars are randomly distributed. Later, in 1903, the Swedish mathematician Filip Lundberg (1876–1965) proposed in his doctoral thesis *"Approximations of the probability function/reinsurance of collective risks"* to model insurance claims using a compound Poisson process. The Danish mathematician, statistician, and engineer AK Erlang (1878–1929) developed in his work entitled by *"The theory of probability and telephone conversations"* a mathematical model to determine the number of incoming telephone calls in a finite time interval. Erlang assumed that the number of incoming telephone calls in each time interval was independent of each other. As part of this study, Erlang determined that the Poisson distribution is a limited form of the binomial distribution.

The New Zealand physicist Ernest Rutherford (1871–1937) and the German physicist Hans Geiger (1882–1945), when analyzing their experimental results on the counting of alpha particles, obtained as a mathematical model a simple Poison process [1]. The Swedish chemist and Nobel laureate in chemistry in 1926, Theodor Svedberg (1884–1971), proposed a model in which a spatial Poisson point process is the underlying process to study how plants are distributed in plant communities [2]. The theoretical development of the Poisson process had the contribution of several of the most influential mathematicians of the 20th century, such as the Russian mathematician Andrei Kolmogorov (1903–1987), American mathematician William Feller (1906–1970) and the Soviet mathematician Aleksandr Khinchin (1894–1959).

© The Author(s), under exclusive license to Springer Nature Switzerland AG 2023   37
L. Blanco-Castañeda and V. Arunachalam, *Applied Stochastic Modeling*, Synthesis
Lectures on Mathematics & Statistics, https://doi.org/10.1007/978-3-031-31282-3_2

2.1   Poisson Processes

In this section, we collect some basic definitions and properties of the Poisson process, which is a suitable mathematical model for describing situations in which events occur randomly over time. For example, the number of flight arrivals at the airport, the number of people who entered a store on or before time t, and the number of claims incurred in the insurance company time t. The Poisson process $\{N_t; t \geq 0\}$ is characterized by the number of events in the interval of time $(0, t]$ for $t > 0$.

Definition 2.1 The stochastic process $\{N_t; t \geq 0\}$ is a *homogeneous Poisson process* or simply a *Poisson process* with parameter $\lambda > 0$, then if it satisfies the following conditions:

1. $N_0 = 0$,
2. $\{N_t; t \geq 0\}$ has independent and stationary increments.
3. It has unit jumps, i.e.,

$$P(N_h = 1) = \lambda h + o(h)$$
$$P(N_h \geq 2) = o(h)$$

where $N_h := N_{t+h} - N_t$.

Remark 2.1 A function $f(\cdot)$ is said to be $o(h)$ if $lim_{h \to \infty} \frac{f(h)}{h} = 0$, which means the function f decays at a faster rate than h.

Alternatively, we can give the definition of the Poisson process as follows.

Definition 2.2 A process $\{N_t; t \geq 0\}$ is a Poisson process with parameter $\lambda > 0$, then if it satisfies the following conditions:

1 $N_0 = 0$.
2 $\{N_t; t \geq 0\}$ has independent and stationary increments.
3. For all $t \geq 0$, N_t has a Poisson distribution with parameter λt:

$$P(N_t = n) = \frac{(\lambda t)^n}{n!} e^{-\lambda t}, \quad n = 0, 1, \ldots \tag{2.1}$$

The Definitions 2.1 and 2.2 are equivalent and interested reader may refer to Ross [3]. We can easily verify that the mean and variances of occurrences in an interval $[0, t]$ and is given by

$$E[N_t] = \lambda t \tag{2.2}$$
$$Var[N_t] = \lambda t \tag{2.3}$$

Fig. 2.1 Simulated sample
path of Poisson process with
$\lambda = 2.5$

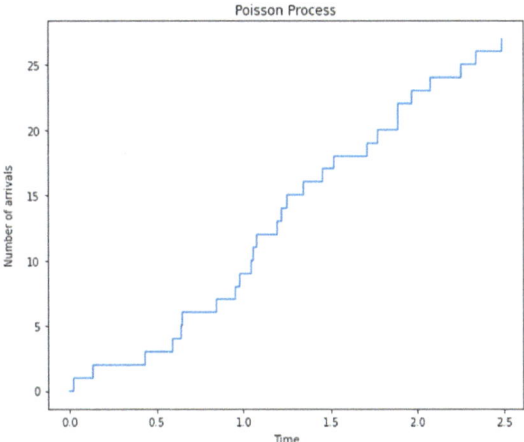

Also, the auto-covariance for $s < t$ is

$$C(N_s, N_t) = \lambda s$$

The simulated sample path of the Poisson process given in Fig. 2.1.

```
def Poisson_Process(t,Lambda):
    k=np.random.poisson(Lambda*t)
    Un=np.random.uniform(0.0, t,k)
    Un.sort()
    N=np.arange(k+1)
    Un=np.insert(Un, 0,0)
    return Un,N
Poisson_Process(2.5,10)
```

Remark 2.2 The Poisson process is a Markov process such that the conditional probabilities are constant and independent of time t.

Let $\{N_t; t \geq 0\}$ be a Poisson process with parameter $\lambda > 0$. If T_n is the time between the $(n-1)$th and nth event, then $\{T_n;\ n = 1, 2, \dots\}$ are the *inter-arrival times* or *holding times* of N_t, and $S_n = \sum_{i=1}^{n} T_i$, for $n \geq 1$ is the *arrival time* of the nth event or the *waiting time* to the nth event.

We are now interested in the distribution of the inter-arrival times. We prove that the inter-arrival time T_i's have an exponential distribution with mean $\frac{1}{\lambda}$. Let T_1 be the time at which the first event occurs. Then

$$P(T_1 > t) = P(N_t = 0) = e^{-\lambda t} .$$

Thus T_1 has an exponential distribution with an expected value $\frac{1}{\lambda}$. Now:

$$\begin{aligned}
P(T_2 > t|T_1 = s) &= P(0 \text{ events in } (s, s + t]|T_1 = s) \\
&= P(0 \text{ events in } (s, s + t]) \quad \text{(independent increments)} \\
&= P(0 \text{ events in } (0, t]) \quad \text{(stationary increments)} \\
&= e^{-\lambda t} .
\end{aligned}$$

Thus T_2 also has an exponential distribution with expected value $\frac{1}{\lambda}$. Note also that T_1 and T_2 are independent, and in general, we have that the inter-arrival times T_n, $n = 1, 2, \ldots$, are independent and identically distributed random variables, each with an expected value $\frac{1}{\lambda}$. We can easily see that the S_n, the arrival time of the nth event, has a gamma(n, λ) distribution, and its probability distribution function of is given by:

$$f_{S_n}(t) = \lambda e^{-\lambda t} \frac{(\lambda t)^{n-1}}{(n-1)!}, \quad t \geq 0. \tag{2.4}$$

And, therefore

$$P\left(S_n \leq t\right) = P\left(N_t \geq n\right) = 1 - \sum_{k=0}^{n-1} \frac{(\lambda t)^k}{k!} \exp\left(-\lambda t\right)$$

Also, the moment generating function of inter-arrival time T_1 is given by

$$\int_0^t e^{-\theta t} \lambda e^{-\lambda t} dt = \frac{\lambda}{\lambda + \theta}$$

and the moment generating function of waiting time S_n is given by

$$\left(\frac{\lambda}{\lambda + \theta}\right)^n . \tag{2.5}$$

Let $\{N_t; t \geq 0\}$ a Poisson process with intensity λ. Suppose that the process is observed until a number fixed n of events. The random variable

$$Z := 2\lambda S_n$$

has a chi-square distribution with $2n$ degrees of freedom.
If $c \in (0, 1)$ and α and β are values such that:

$$P\left(Z \leq \alpha\right) = \frac{c}{2} = P\left(Z \geq \beta\right)$$

then

$$P(\alpha \le Z \le \beta) = 1 - P(Z < \alpha \lor Z > \beta)$$
$$= 1 - [P(Z < \alpha) + P(Z > \beta)]$$
$$= 1 - c.$$

Therefore,

$$P(\alpha \le Z \le \beta) = P\left(\frac{\alpha}{2S_n} \le \lambda \le \frac{\beta}{2S_n}\right) = 1 - c$$

that is,

$$\left[\frac{\alpha}{2S_n}, \frac{\beta}{2S_n}\right] = \left[\frac{\chi^2_{2n,\frac{c}{2}}}{2S_n}, \frac{\chi^2_{2n,1-\frac{c}{2}}}{2S_n}\right]$$

is a confidence interval for λ at the level $(1 - c)$.

Suppose that now exactly one event of a Poisson process occurs during the interval $(0, t]$. Then the conditional distribution of T_1 given that $N_t = 1$ is uniformly distributed over the interval $(0, t]$ is:

$$P(T_1 \le s \mid N_t = 1) = \frac{P(T_1 \le s, N_t = 1)}{P(N_t = 1)}$$
$$= \frac{P(N_s = 1, N_{t-s} = 0)}{P(N_t = 1)}$$
$$= \frac{P(N_s = 1)P(N_{t-s} = 0)}{P(N_t = 1)}$$
$$= \frac{\lambda s e^{-\lambda s} \cdot e^{-\lambda(t-s)}}{\lambda t e^{-\lambda t}}$$
$$= \frac{s}{t}.$$

The following proposition shows that the sum of two independent Poisson processes is also a Poisson process.

Proposition 2.1 *Let $\{N_t; t \ge 0\}$ and $\{\widehat{N}_t; t \ge 0\}$ independent Poisson processes with parameters λ and $\widehat{\lambda}$ respectively. Then $\{X_t : t \ge 0\}$ where $X_t = N_t + \widehat{N}_t$ is a Poisson process with parameter $\lambda + \widehat{\lambda}$.*

Proof Let us see that $\{X_t; t \ge 0\}$ has independent increments.

Suppose $t_0 < t_1 < \cdots < t_k$ and let $n_1, n_2, \ldots, n_k, m_1, m_2, \ldots, m_k \in \mathbb{N}$. Then:

$$P(N_{t_k} - N_{t_{k-1}} = n_k, \widehat{N}_{t_k} - \widehat{N}_{t_{k-1}} = m_k, \ldots, N_{t_1} - N_{t_0} = n_1, \widehat{N}_{t_1} - \widehat{N}_{t_0} = m_1)$$
$$= P(N_{t_k} - N_{t_{k-1}} = n_k, \ldots, N_{t_1} - N_{t_0} = n_1, \widehat{N}_{t_k} - \widehat{N}_{t_{k-1}} = m_k, \ldots,$$
$$\widehat{N}_{t_1} - \widehat{N}_{t_0} = m_1)$$
$$= P\left(N_{t_k} - N_{t_{k-1}} = n_k, \ldots, N_{t_1} - N_{t_0} = n_1\right)$$
$$\cdot P(\widehat{N}_{t_k} - \widehat{N}_{t_{k-1}} = m_k, \ldots, \widehat{N}_{t_1} - \widehat{N}_{t_0} = m_1)$$
$$= \prod_{j=1}^{k} P\left(N_{t_j} - N_{t_{j-1}} = n_j\right) P\left(\widehat{N}_{t_j} - \widehat{N}_{t_{j-1}} = m_j\right)$$

Consequently, the random vectors

$$\left(N_{t_1} - N_{t_0}, \widehat{N}_{t_1} - \widehat{N}_{t_0}\right), \left(N_{t_2} - N_{t_1}, \widehat{N}_{t_2} - \widehat{N}_{t_1}\right), \ldots,$$
$$\left(N_{t_k} - N_{t_{k-1}}, \widehat{N}_{t_k} - \widehat{N}_{t_{k-1}}\right)$$

are independent. As the map $g : \mathbb{R}^2 \longrightarrow \mathbb{R}$ given by $g(x, y) = x + y$ is measurable and continuous then the random variables

$$X_{t_1} - X_{t_0} = \left(N_{t_1} - N_{t_0}\right) + \left(\widehat{N}_{t_1} - \widehat{N}_{t_0}\right), \ldots,$$
$$X_{t_k} - X_{t_{k-1}} = \left(N_{t_k} - N_{t_{k-1}}\right) + \left(\widehat{N}_{t_k} - \widehat{N}_{t_{k-1}}\right)$$

are independent.

Since $N_t \overset{d}{=} \mathcal{P}(\lambda t)$, $\widehat{N}_t \overset{d}{=} \mathcal{P}(\widehat{\lambda} t)$ and the variables N_t and $\widehat{N}_t$ are independent, then $X_t \overset{d}{=} \mathcal{P}\left(\left(\lambda + \widehat{\lambda}\right) t\right)$.

Theorem 2.1 *Let $\{N_t; t \geq 0\}$ be a Poisson process with parameter λ. Then the joint conditional density of $T_1, T_2, \ldots, T_n$ given $N_t = n$ is*

$$f_{T_1, T_2, \ldots, T_n | N_t = n}(t_1, t_2, \ldots, t_n) = \begin{cases} \frac{n!}{t^n} & \text{if } 0 < t_1 < t_2 < \cdots < t_n < t \\ 0 & \text{other cases} . \end{cases}$$

The Poisson process is a purely random process since the occurrences are equally likely to occur anywhere in the interval $[0, t]$, and the event occurs at times $t_1, t_2, \ldots t_n$ are randomly distributed over the interval $[0, t]$. The relationship between the Poisson process and the uniform distribution has stated in the following proposition.

Proposition 2.2 *Let $\{N_t; t \geq 0\}$ a Poisson process with intensity λ. Under the condition $N_t = n$ we have that the time vector real $(S_1, S_2, \ldots, S_n)$ at which the events have the same distribution as the vector of statistics of order $\left(U_{(1)}, U_{(2)}, \ldots, U_{(n)}\right)$ from a random sample $U_1, U_2, \ldots, U_n$ of the uniform distribution over the interval $(0, t]$.*

Proof It is known that the general formula of the joint probability density function of order statistics $U_{(1)}, U_{(2)}, \ldots, U_{(n)}$ of a random sample $U_1, U_2, \ldots, U_n$ of a density function f is given by

$$f_{U_{(1)},U_{(2)},\ldots,U_{(n)}} (x_1, x_2, \ldots, x_n) = n! f(x_1) \ldots f(x_n)$$

with $x_1 < x_2 < \cdots < x_n$. When the common distribution is uniform over the interval $(0, t]$ is obtained. in particular

$$f_{U_{(1)},U_{(2)},\ldots,U_{(n)}} (t_1, t_2, \ldots, t_n) = \begin{cases} \frac{n!}{t^n} & \text{if } t_1 < t_2 < \cdots < t_n \\ 0 & \text{otherwise} \end{cases} \quad (*)$$

We must therefore prove that the joint distribution of $S_1, S_2, \ldots, S_n$ is given above $(*)$. We have the conditional density function of the vector $(S_1, S_2, \ldots, S_n)$ given that $N_t = n$ is equal to

$$f_{(S_1,S_2,\ldots,S_n)|N_t} (t_1, t_2, \ldots, t_n \mid n)$$

$$= \frac{\partial^n}{\partial t_1 \ldots \partial t_n} P(S_1 \leq t_1, \ldots, S_n \leq t_n \mid N_t = n)$$

$$= \frac{\partial^n}{\partial t_1 \ldots \partial t_n} P(N_{t_1} \geq 1, \ldots, N_{t_n} \geq n \mid N_t = n)$$

$$= \frac{\partial^n}{\partial t_1 \ldots \partial t_n} P(N_t - N_{t_n} = 0, N_{t_n} - N_{t_{n-1}} = 1, \ldots,$$

$$N_{t_2} - N_{t_1} = 1, N_{t_1} = 1 \mid N_t = n)$$

$$= \frac{\partial^n}{\partial t_1 \ldots \partial t_n} \frac{\left(\exp(-\lambda(t - t_n)) \prod_{j=1}^{n} \lambda(t_j - t_{j-1}) \exp(-\lambda(t_j - t_{j-1})) \right)}{\frac{(\lambda t)^n}{n!} \exp(-\lambda t)}$$

$$= \frac{\partial^n}{\partial t_1 \ldots \partial t_n} \frac{n! \prod_{j=1}^{n} (t_j - t_{j-1})}{t^n}$$

$$= \frac{n!}{t^n}$$

with $0 = t_0 < t_1 < t_2 < \cdots < t_n < t$.

Suppose that a certain situation has been observed that occurs randomly, e.g. customer arrivals at a bank, during a given period of time. If the process that denotes the number of events occurred is Poisson and if up to the instant t the occurrence of n events has been observed then, according to the previous result, for moderately large values of n it must be satisfied that:

$$\frac{nt}{2} - 1.96\sqrt{\frac{nt^2}{12}} \le \sum_{j=1}^{n} S_j \le \frac{nt}{2} + 1.96\sqrt{\frac{nt^2}{12}}$$

where $(S_1, S_2, \ldots, S_n)$ is the vector of real times in which the events occur. For example, if during a period of 10 min 20 customers have arrived in a waiting queue and if the actual arrival times of the customers, measured in minutes and counted from instant 0 in that the observation began, were

$$0.2, 0.25, 0.38, 0.43, 0.45$$
$$1.32, 1.56, 2.02, 2.89, 3.92$$
$$4.45, 4.78, 5.34, 5.76, 6.48$$
$$7.23, 7.82, 8.67, 9.21, 9.77$$

then we have in this case that the sum of the real times of occurrence of the events is equal to 82. 93. Since

$$\frac{nt}{2} - 1.96\sqrt{\frac{nt^2}{12}} = \frac{20 \times 10}{2} - 1.96\sqrt{\frac{20 \times 100}{12}} = 74.697$$

and

$$\frac{nt}{2} + 1.96\sqrt{\frac{nt^2}{12}} = \frac{20 \times 10}{2} + 1.96\sqrt{\frac{20 \times 100}{12}} = 125.367$$

Then we can accept the hypothesis that the events occurred according to a Poisson process with a significance level of 95%.

Proposition 2.3 Let $\{N_t; t \ge 0\}$ a Poisson process with intensity λ and suppose that at the moment s of an event is classified as type I with probability $p(s)$ or type II with probability $(1 - p(s))$. If $N_1(t)$ and $N_2(t)$ are the random variables that denote, respectively, the number of events of type I and of type II that occur in the interval $(0, t]$ then $N_1(t)$ and $N_2(t)$ are independent and we have that $N_1(t) \overset{d}{=} \mathcal{P}(\lambda pt)$ and $N_2(t) \overset{d}{=} \mathcal{P}(\lambda(1 - p)t)$ where

$$p := \frac{1}{t}\int_0^t p(s)\, ds$$

Proof Suppose that in the interval $(0, t]$ n events of type I and m events of type II have occurred. We then have that:

$$P(N_1(t) = n, N_2(t) = m) = \sum_k P(N_1(t) = n, N_2(t) = m \mid N_t = k) P(N_t = k)$$

$$= P(N_1(t) = n, N_2(t) = m \mid N_t = n + m) P(N_t = n + m)$$

$$= P(N_1(t) = n, N_2(t) = m \mid N_t = n + m) \frac{(\lambda t)^{n+m}}{(n+m)!} \exp(-\lambda t)$$

On the other hand, the probability that $N_1(t) = n$ and $N_2(t) = m$ given that $N_t = n + m$ is equal to the probability of getting n successes and m failures in a Bernoulli sequence of length $(n + m)$ with a probability of success p equal to the probability that the event is classified as type I, and with a probability of failure equal to the probability that the event is classified as type II. Suppose that an arbitrary event has occurred in the time interval $(0, t]$ and let T the random variable denoting the instant at which the event occurred and it is known that under the condition $N_t = 1$, the random variable T has a uniform distribution over the interval $(0, t]$. Therefore,

$$p = P \text{ ("the event is of type I")}$$
$$= E\left(E\left(X_{\{\text{type I}\}}\right) \mid T\right)$$
$$= \int_0^t p(s) \frac{1}{t} ds$$
$$= \frac{1}{t} \int_0^t p(s) ds$$

Consequently

$$P(N_1(t) = n, N_2(t) = m \mid N_t = n + m) = \binom{n+m}{n} p^n q^m$$

Finally, we concluded that

$$P(N_1(t) = n, N_2(t) = m)$$
$$= \exp(-\lambda t p) \frac{(\lambda t p)^n}{n!} \exp(-\lambda t (1 - p)) \frac{(\lambda t (1 - p))^m}{m!}$$

Example 2.1 Suppose that calls enter a customer service center according to a Poisson process of intensity λ. As soon as the call arrive in, it is answered. Assume that the call duration times are Independent and identically distributed random variables with a given distribution F. What is the distribution of the number of calls that have been completely answered until time t?, What is the distribution of the number of calls that have not yet completely answered until time t?

Solution: We classify calls into two types:
   Type I: calls that have been answered until time t.
   Type II: calls that have not yet been completely answered until time t.

If a call is received at time s, then it would be type I if its duration is less than or equal to $(t - s)$ which happens with probability $F(t - s)$ and is type II if its duration is greater than $(t - s)$, which occurs with probability $1 - F(t - s)$. Therefore, if

$$N_1(t) := \text{"number of type I calls"}$$
$$N_2(t) := \text{"number of type II calls"}$$

then

$$N_1(t) \stackrel{d}{=} \mathcal{P}(\lambda t p)$$

and

$$N_2(t) \stackrel{d}{=} \mathcal{P}(\lambda t (1 - p))$$

where

$$p := \frac{1}{t} \int_0^t F(t - s)\, ds$$

The following result shows that homogeneous Poisson processes can be decomposed into independent Poisson processes.

Theorem 2.2 *Let $\{N_t; t \geq 0\}$ a Poisson process with intensity λ. Assume that $\{X_n; n \geq 0\}$ n is a sequence of independent and equally distributed random variables with Bernoulli distribution of parameter $0 < p < 1$. We have the processes $\left\{N_t^{(1)}; t \geq 0\right\}$ and $\left\{N_t^{(2)}; t \geq 0\right\}$ with*

$$N_t^{(1)} := \sum_{k=0}^{N_t} X_k$$

and

$$N_t^{(2)} := \sum_{k=0}^{N_t} (1 - X_k)$$

are homogeneous and independent Poisson processes with intensities λp and $\lambda(1 - p)$ respectively.

Proof We have that

$$N_0^{(1)} = 0 = N_0^{(2)}$$

On the other hand, if $t_0 < t_1 < \cdots < t_n$ then for $i = 1, 2$ we have that the random variables

$$N_{t_0}^{(i)}, \ N_{t_1}^{(i)} - N_{t_0}^{(i)}, \ N_{t_2}^{(i)} - N_{t_1}^{(i)}, \ \ldots, \ N_{t_n}^{(i)} - N_{t_{n-1}}^{(i)}$$

are independent since, by hypothesis, the random variables $\{X_n; n \geq 0\}$ are independent.

We have

$$P\left(N_t^{(1)} = k\right) = P\left(\sum_{k=0}^{N_t} X_k = k\right)$$

$$= \sum_{n=0}^{\infty} P\left(\sum_{k=0}^{n} X_k = k \mid N_t = n\right) P\left(N_t = n\right)$$

$$= \sum_{n=k}^{\infty} \binom{n}{k} p^k (1-p)^{n-k} \exp(-\lambda t) \frac{(\lambda t)^n}{n!}$$

$$= \exp(-\lambda t) \frac{(\lambda t p)^k}{k!} \sum_{j=0}^{\infty} \frac{(\lambda t (1-p))^j}{j!}$$

$$= \exp(-\lambda t) \frac{(\lambda t p)^k}{k!} \exp(\lambda t (1-p))$$

$$= \exp(-\lambda t p) \frac{(\lambda t p)^k}{k!}$$

and

$$P\left(N_t^{(2)} = k\right) = P\left(\sum_{k=0}^{N_t} (1 - X_k) = k\right)$$

$$= \sum_{n=0}^{\infty} P\left(\sum_{k=0}^{n} (1 - X_k) = k \mid N_t = n\right) P\left(N_t = n\right)$$

$$= \sum_{n=0}^{\infty}\sum_{n=k}^{\infty} \binom{n}{k} (1-p)^k \, p^{n-k} \exp(-\lambda t) \frac{(\lambda t)^n}{n!}$$

$$= \exp(-\lambda t) \frac{(\lambda t (1-p))^k}{k!} \sum_{j=0}^{\infty} \frac{(\lambda t p)^j}{j!}$$

$$= \exp(-\lambda t) \frac{(\lambda t (1-p))^k}{k!} \exp(\lambda t p)$$

$$= \exp(-\lambda (1-p) t) \frac{(\lambda t (1-p))^k}{k!}.$$

Now we check that the processes are independent.

$$P\left(N_t^{(1)} = k, N_t^{(2)} = l\right) = P\left(N_t^{(1)} = k, N_t = k + l\right)$$

$$= P\left(N_t^{(1)} = k \mid N_t = k + l\right) P\left(N_t = k + l\right)$$

$$= \binom{k+l}{k} p^k (1 - p)^l \exp(-\lambda t) \frac{(\lambda t)^{k+l}}{(k+l)!}$$

$$= \frac{p^k \exp(-\lambda p t) (1 - p)^l \exp(-\lambda (1 - p) t)}{k!} \frac{}{l!}$$

$$= P\left(N_t^{(1)} = k\right) P\left(N_t^{(2)} = l\right).$$

Remark 2.3 Suppose that the events of a homogeneous Poisson process $\{N_t; t \geq 0\}$ with intensity λ, can be classified in m disjoint classes of events with occurrence probabilities $0 < p_i < 1$ with $i = 1, 2, \ldots, m$ and $\sum_{i=1}^{m} p_i = 1$ then if we define, for $i = 1, 2, \ldots, m$,

$N_t^{(i)} :=$ "number of events of type i", we have that the processes $\left\{N_t^{(i)}; t \geq 0\right\}$ are independent Poisson processes with intensities $\lambda_i = \lambda p_i$ respectively.

Example 2.2 Suppose potential adopters arrive at a pet adoption center according to a Poisson process with intensity $\lambda = 20$ per day. At the adoption center there are cats and dogs to adopt. 40% of the people who enter the center are interested in adopting a cat and the remaining 60% prefer to adopt a dog. Of the people who want to adopt a dog, 50% prefer small size dogs, 40% medium size dogs and 10% big size dogs. It is also known that, of the people who are interested in adopting a small size dog, 30% do so, of those who are interested in a medium size dog, 20 indeed does, of those who are interested in a big size dog, 10% indeed adopts it. In relation to cats, it is known that 10% of those interested in adopting them do so. What is the probability that, in one day, the center manages to find an adoptive family for at least three animals?

Solution:

We define:

$N_t :=$ "number of people entering the adoption center in $(0, t]$"

$N_t^D :=$ "number of people interested in adopting a dog in $(0, t]$"

$N_t^G :=$ "number of people interested in adopting a cat in $(0, t]$"

We have that $\{N_t^D; t \geq 0\}$ and $\{N_t^G; t \geq 0\}$ are independent Poisson processes with intensities $\lambda_D = 12$ and $\lambda_G = 8.0$ respectively. In addition, if we consider:

$N_t^{D,1} :=$ "number of people interested in adopting a big size dog in $(0, t]$"

$N_t^{D,2} :=$ "number of people interested in adopting a medium size dog in $(0, t]$"

$N_t^{D,3} :=$ "number of people interested in adopting a small size dog in $(0, t]$"

We see that $\left(N_t^{D,i}\right) t \geq 0$ with $i = 1, 2, 3$ are independent Poisson processes with intensities $\lambda_{D,1} = 1.2$, $\lambda_{D,2} = 4.8$ and $\lambda_{D,3} = 6$ respectively. Now, we are interested the number of people interested in adopting an animal, and if

$N_t^{*D,1} :=$ "number of people who are interested in adopting a big size dog
  and actually do so in $(0, t]$"

$N_t^{*D,2} :=$ "number of people who are interested in adopting a medium size dog
  and actually do so in $(0, t]$"

$N_t^{*D,3} :=$ "number of people interested in adopting a small size dog
  and actually do so in $(0, t]$"

$N_t^{*G} :=$ "number of people who are interested in adopting a cat
  and actually do so in $(0, t]$"

We know that there are turn out to be independent Poisson processes with intensities $\lambda_{D,1}^* = 0.12$, $\lambda_{D,2}^* = 0.96$ and $\lambda_{D,3}^* = 1.8$ and $\lambda_G^* = 0.8$ respectively.
    Therefore, we have

$N_t^* :=$ "number of people who are interested in adopting an animal
  and actually do so in $(0, t]$"

hence $\left(N_t^*\right) t \geq 0$ is a Poisson process with intensity $\lambda^* = 3.68$. We get

$$P\left(N_{t+1}^* - N_t^* \geq 3\right) = 1 - \sum_{k=0}^{2} \frac{(3.68)^k}{k!} \exp\left(-3.68\right)$$

$$= 1 - \left[\exp\left(-3.68\right) + 3.68 \times \exp\left(-3.68\right) + \frac{1}{2}(3.68)^2 \exp\left(-3.68\right)\right]$$

$$= 1 - 0.288\,83$$

$$= 0.771117.$$

Proposition 2.4 *Let $\{N_t; t \geq 0\}$ be a Poisson process with intensity λ. Let $s, t \in [0, \infty)$ with $s < t$ and $k, n \in \mathbb{N}$ con $k \leq n$. we have that:*

$$P\left(N_s = k \mid N_t = n\right) = \binom{n}{k}\left(\frac{s}{t}\right)^k\left(1 - \frac{s}{t}\right)^{n-k}$$

Proof From the properties of the Poisson process, we have:

$$
\begin{aligned}
P\left(N_s = k \mid N_t = n\right) &= \frac{P\left(N_s = k,\ N_t = n\right)}{P\left(N_t = n\right)} \\
&= \frac{P\left(N_s = k,\ N_{t-s} = n - k\right)}{P\left(N_t = n\right)} \\
&= \frac{P\left(N_s = k\right) P\left(N_{t-s} = n - k\right)}{P\left(N_t = n\right)} \\
&= \frac{\exp\left(-\lambda s\right) \frac{(\lambda s)^k}{k!} \exp\left(-\lambda\left(t - s\right)\right) \frac{[\lambda(t-s)]^{n-k}}{(n-k)!}}{\exp\left(-\lambda t\right) \frac{(\lambda t)^n}{n!}} \\
&= \binom{n}{k}\left(\frac{s}{t}\right)^k\left(\frac{t - s}{t}\right)^{n-k}.
\end{aligned}
$$

Example 2.3 Suppose that customers arrive at a store according to a Poisson process with intensity $\lambda = 10$ per hour. If the store opens at 8 : 00 a.m. and by 10 : 00 a.m. there are 15 customers have arrived. What is the probability that between 8 : 00 a.m. and 9 : 00 a.m. exactly 10 customers have to arrived?

Solution: Let $N(t) :=$ "number of customers that arrive at the store in the interval $(0, t]$" Here time t is measured in hours. According to the previous theorem, we have

$$P\left(N\left(1\right) = 10 \mid N\left(2\right) = 15\right) = \binom{15}{10}\left(\frac{1}{2}\right)^{10}\left(1 - \frac{1}{2}\right)^5$$

$$= 9.\,164\,4 \times 10^{-2}.$$

Theorem 2.3 *Let $\{N_t;\ t \geq 0\}$ a Poisson process with intensity λ. If the process is observed during a time interval $(0, T]$ fixed then $\widehat{\lambda} = \frac{N_T}{T}$ is the maximum likelihood estimator of the parameter λ.*

Proof We have

$$
\begin{aligned}
E\left(\widehat{\lambda}\right) &= \frac{E\left(N_T\right)}{T} \\
&= \frac{\lambda T}{T} \\
&= \lambda
\end{aligned}
$$

and

$$Var\left(\widehat{\lambda}\right) = \frac{Var\,(N_T)}{T^2}$$

$$= \frac{\lambda T}{T^2}$$

$$= \frac{\lambda}{T}$$

Now

$$I := E\left(\left[\frac{d}{d\lambda}\log\left(f_{N_T}\left(\lambda\right)\right)\right]^2\right)$$

$$= E\left(\left[\frac{d}{d\lambda}\log\left(\frac{(\lambda T)^{N_T}}{N_T!}\exp\left(-\lambda T\right)\right)\right]^2\right)$$

$$= E\left(\left[\frac{d}{d\lambda}\left(\log\left(\frac{(\lambda T)^{N_T}}{N_T!}\right) + (-\lambda T)\right)\right]^2\right)$$

$$= E\left(\left[\frac{d}{d\lambda}\left(N_T\log\left(\lambda T\right) - \log\left(N_T!\right) - \lambda T\right)\right]^2\right)$$

$$= E\left(\frac{N_T}{\lambda} - T\right)^2$$

$$= E\left(\frac{N_T^2}{\lambda^2} - 2T\frac{N_T}{\lambda} + T^2\right)$$

$$= \frac{1}{\lambda^2}\left(\lambda T + \lambda^2 T^2\right) - 2T\frac{\lambda T}{\lambda} + T^2$$

$$= \frac{T}{\lambda} + T^2 - 2T^2 + T^2$$

$$= \frac{T}{\lambda}$$

which matches $Var\left(\widehat{\lambda}\right)$.

Example 2.4 In the case of the example given in Example 2.3, the maximum likelihood estimator of the parameter λ is given for:

$$\widehat{\lambda} = \frac{20}{10} = 2$$

That is, on average, $\frac{1}{2}$ min elapses between successive customer arrivals.

2.2   Non-homogeneous Poisson Process

In this section we generalize the Poisson process by assuming that it has a non-stationary increment, or a time-dependent intensity function $\lambda(t)$, we call the process $\{N_t; t \geq 0\}$ a non-homogeneous Poisson process and also called a non-homogeneous Poisson process. For example, the number of customers who arrive at a store follows a certain duration time in the evening compared to the morning time. The Poisson process with parameter λ depends on t, so $\lambda(t)$ is called the intensity function of the process, which is a non-negative and integrable function. We have the following definition:

Definition 2.3 The stochastic processes $\{N_t; t \geq 0\}$ is a *non-homogeneous Poisson process* with intensity function $\lambda(t), t \geq 0$, if:

1. $N_0 = 0$.
2. $\{N_t; t \geq 0\}$ has independent increments.
3. For $0 \leq s < t$, the random variable $N_t - N_s$ has a Poisson distribution with parameter $\int_s^t \lambda(u)du$. That is,

$$P(N_t - N_s = k) = \frac{\left(\int_s^t \lambda(u)du\right)^k}{k!} e^{-\int_s^t \lambda(u)du}$$

for $k = 0, 1, 2, \ldots$.

Remark 2.4 We define the mean value function

$$m(t) = \int_0^t \lambda(u)du.$$

We give the realizations of an non-homogeneous Poisson process for the intensity functions $\lambda_1(t) = e^{-\frac{t}{5}} + \frac{1}{5}$, $\lambda_2(t) = \sin t + 1$ and $\lambda_3(t) = \frac{1}{2}\sqrt{\frac{t}{2}}$. These graphs were generated in Python with the following code:

```
import scipy.integrate as integrate
lista=[]
for i in range(1,5001):
    integral=integrate.quad(lambda x: np.sin(x)+1, (i-1)*0.01, i*0.01)
    k=np.random.poisson(integral[0])
    lista.append(k)
Final=np.cumsum(lista)
x=np.arange(len(Final))
plt.step(x,Final,where='post' )
plt.xlabel("t")
plt.ylabel("Nta")
plt.show()
```

Example 2.5 A customer service office of a certain entity begins its working day at 9:00 and ends at 18:00. Suppose that from 9:00 a.m. to 11:00 a.m. service users arrive at a constant rate of 10 users per hour, from 11:00 a.m. to 3:00 p.m. the user arrival rate grows linearly from 10 users at 11:00 a.m. to 8:00 p.m. and from 3:00 p.m. to 6:00 p.m., the arrival rate of users decreases linearly from 20 to 8 users. What is the probability that no user arrives between 9:30 a.m. and 11:30 a.m.?

Solution: Let N_t be the number of users who arrive at the customer service office in the time interval $(0, t]$. A suitable model to describe the situation described in this example is an non-homogeneous Poisson process with an intensity function given by (Figs. 2.2, 2.3 and 2.4):

Fig. 2.2 Poisson process with $\lambda_1(t) = e^{-\frac{t}{5}} + \frac{1}{5}$

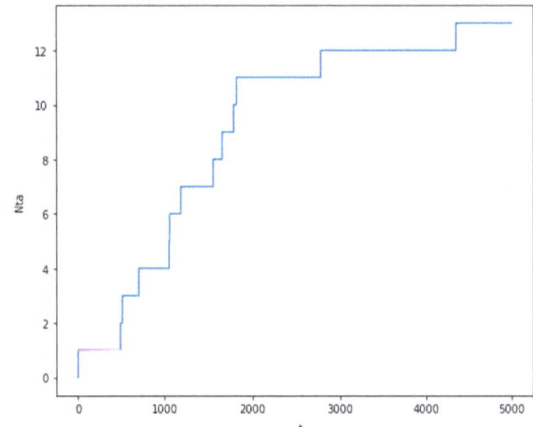

Fig. 2.3 Poisson process with $\lambda_2(t) = \sin t + 1$

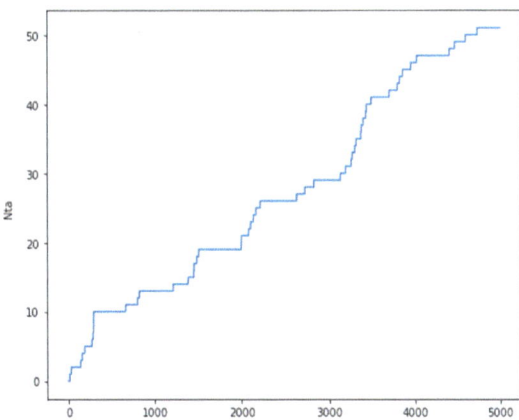

Fig. 2.4 Poisson process with
$\lambda_3(t) = \frac{1}{2}\sqrt{\frac{t}{2}}$

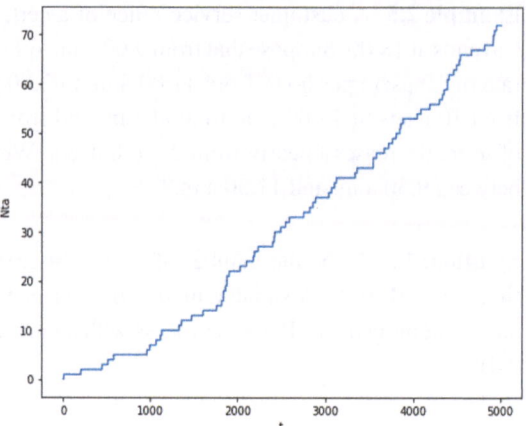

$$\lambda(t) = \begin{cases} 10 & \text{if } 0 \le t < 2 \\ \frac{5}{2}t + 5 & \text{if } 2 \le t < 6 \\ -4t + 44 & \text{if } 6 \le t \le 9 \end{cases}$$

and we assume $\lambda(t) = \lambda(9 - t)$ for $t > 9$. In this case,

$$(N_{2.5} - N_{0.5}) \stackrel{d}{=} \mathcal{P}(m(2.5) - m(0.5))$$

where

$$m(2.5) = \int_0^2 10ds + \int_2^{2.5} \left(\frac{5}{2}t + 5\right) dt = 25.31$$

$$m(0.5) = \int_0^{0.5} 10ds = 5$$

Therefore,

$$P(N_{2.5} - N_{0.5} = 0) = \exp(-20.313)$$
$$= 1.5072 \times 10^{-9}.$$

Remark 2.5 Let $\{N_t; t \ge 0\}$ an non-homogeneous Poisson process with intensity function $\lambda(t)$. Suppose that S_n denotes the elapsed time from the moment the observation begins to the moment the nth event occurs.

Consider

$$P(t < S_n < t + h) = P(N_t = n - 1, \text{"an event occurs in } (t, t + h)\text{"})$$

$$= P(N_t = n - 1) P(N_{t+h} - N_t = 1)$$

$$= \exp(-m(t)) \frac{(m(t))^{n-1}}{(n-1)!}$$

$$\times \exp(m(t+h) - m(t))(m(t+h) - m(t))$$

Dividing by h and making h tend towards zero, we obtain:

$$f_{S_n}(t) = \exp(-m(t)) \times \frac{(m(t))^{n-1}}{(n-1)!} \lambda(t)$$

Suppose that T_1 and T_2 denote the times at which the first and second events occurs. It's clear that:

$$f_{T_1}(t) = \lambda(t) \exp(-m(t))$$

And

$$P(T_2 > t \mid T_1 = s) = P(\text{occurs 0 event in } (s, s + t] \mid T_1 = s)$$

$$= P(\text{occurs 0 events in } (s, s + t])$$

$$= \exp(-[m(s + t) - m(s)])$$

Therefore, the random variables T_1 and T_2 are not independent. Furthermore,

$$P(T_2 > t) = \int_0^\infty P(T_2 > t \mid T_1 = s) f_{T_1}(s) \, ds$$

$$= \int_0^\infty \lambda(s) \exp(-m(s)) \times \exp(-[m(s + t) - m(s)]) \, ds$$

$$= \int_0^\infty \lambda(s) \exp(-m(t + s)) \, ds$$

and hence

$$f_{T_2}(t) = -\frac{d}{dt} P(T_2 > t)$$

$$= \int_0^\infty \lambda(s) \lambda(t + s) \exp(-m(t + s)) \, ds$$

We can see that the random variables T_1 and T_2 do not have the same distribution.

Theorem 2.4 *Let $\{N_t; t \geq 0\}$ an non-homogeneous Poisson process with intensity function $\lambda(t)$ and if mean value function $m(t)$ then under the condition $N_t = n$ we have that the time vector real $(S_1, S_2, \ldots, S_n)$ in which the events occur has the same distribution as the vector statistics of order $(U_{(1)}, U_{(2)}, \ldots, U_{(n)})$ of a random sample of random variables*

$U_1, U_2, \ldots, U_n$ with probability density function is given by:

$$f_{U_k}(s) = \begin{cases} \frac{\lambda(s)}{m(t)} & \text{if } 0 \leq s \leq t \\ 0 & \text{otherwise} \end{cases}$$

with $k = 1, 2, \ldots, n$.

Remark 2.6 From the previous theorem, we obtain that

$$f_{(U_{(1)}, U_{(2)}, \ldots, U_{(n)})}(t_1, t_2, \ldots, t_n) = n! \, (m(t))^{-n} \prod_{k=1}^{n} \lambda(t_k)$$

Proposition 2.5 *Let $\{N_t; t \geq 0\}$ an non-homogeneous Poisson process with mean value function $m(t)$ then the process*

$$\left(N_t^*\right)_{t \geq 0}$$

where

$$N_t^* = N_{[m(t)]^{-1}}$$

with

$$m^{-1}(t) := \inf \{s > 0 \mid m(s) \geq t\}$$

is a Poisson process with intensity $\lambda = 1$.

Proof Let us see that the process $\left(N_t^*\right) t \geq 0$ has independent and stationary increments and for $s < t$ holds:

$$N_t^* - N_s^* \stackrel{d}{=} \mathcal{P}(t - s)$$

Suppose that $0 \leq t_1 < t_2 < \cdots < t_n$ then it is true that:

$$0 \leq m^{-1}(t_1) < m^{-1}(t_2) < \cdots < m^{-1}(t_n)$$

since the function $m(t)$ is increasing. Therefore, the random variables

$$N_{t_1}^*, \ N_{t_2}^* - N_{t_1}^*, \ldots, N_{t_n}^* - N_{t_{n-1}}^*$$

are independent and it also holds that, for $s < t$,

$$N_t^* - N_s^* \stackrel{d}{=} \mathcal{P}\left(m\left(m^{-1}(t)\right) - m\left(m^{-1}(s)\right)\right) = \mathcal{P}(t - s)$$

since $m^{-1}(t)$ is the right inverse of $m(t)$.

According to [4], the following are two non-parametric estimators of the intensity function.

Proposition 2.6 *Let* $\{N_t; t \geq 0\}$ *be a non-homogeneous be an Poisson process with intensity function* $\lambda(t)$.

a. *Suppose that the process is observed in the interval* $[0, T]$ *and this can be divided into* k *intervals (disjoints) of the form* $[0, t_1], (t_1, t_2], \ldots, (t_{k-1}, T]$ *then*

$$\widehat{\lambda}_t^H = \sum_{m=1}^{k} \frac{N_{t_m} - N_{t_{m-1}}}{t_m - t_{m-1}} X_{(t_{m-1}, t_m]}(t)$$

is the histogram estimator of the intensity function $\lambda(t)$.

b. *If the observation interval* $(0, T]$ *cannot be divided into disjoint intervals then for* $\delta > 0$ *fixed we have*

$$\widehat{\lambda}_t = \begin{cases} \frac{N_{t+\delta}}{t+\delta} & \text{if} \quad 0 \leq t < \delta \\ \frac{N_{t+\delta} - N_{t-\delta}}{2\delta} & \text{if} \quad \delta \leq t \leq t - \delta \\ \frac{N_T - N_{t-\delta}}{T - t + \delta} & \text{if} \quad T - \delta \leq t \leq T \end{cases}$$

This estimator is known as the moving mean estimator.

Suppose that we have data that comes from a non-homogeneous Poisson process $\{N_t; t \geq 0\}$ and that we want to find the intensity function that gave rise to that data set. One way to do this is to use the principle of maximum likelihood to find the intensity function $\lambda(t)$ that maximizes the probability of occurrences of that data set. If in the time interval $(0, T]$ n process events have occurred at times $t_1 < t_2 < \cdots < t_n$ then the likelihood function for the sample $\pi = \{t_1, t_2, \ldots t_n\}$ is given for:

$$L(\lambda, \pi) = \exp(-m(T)) \frac{[m(T)]^n}{n!} n! (m(T))^{-n} \prod_{k=1}^{n} \lambda(t_k)$$

$$= \exp(-m(T)) \prod_{k=1}^{n} \lambda(t_k)$$

$$= \exp\left(-\int_0^T \lambda(t) \, dt\right) \prod_{k=1}^{n} \lambda(t_k)$$

and hence

$$l(\lambda) = -\int_0^T \lambda(t)\,dt + \sum_{k=1}^n \log \lambda(t_k)$$

Trying to find the intensity function $\lambda(t)$ is generally not an easy task. Drazek [5] considers the following families of intensity functions for the Poisson processes

1. Polynomials:

$$\lambda(t) = a_0 + a_1 t + \cdots + a_n t^n$$

with $\lambda(t) \geq 0$ for all t.
2. Fourier Series:

$$\lambda(t) = a_0 + \sum_{j=1}^n \left[b_j \sin\left(2\pi \frac{t}{T} f_j\right) + c_j \cos\left(2\pi \frac{t}{T} f_j\right) \right]$$

with $\lambda(t) \geq 0$ for all t.
3. Exponential polynomials:

$$\lambda(t) = \exp\left(a_0 + a_1 t + \cdots + a_n t^n\right)$$

4. Exponential Fourier series:

$$\lambda(t) = \exp\left(a_0 + \sum_{j=1}^n \left[b_j \sin\left(2\pi \frac{t}{T} f_j\right) + c_j \cos\left(2\pi \frac{t}{T} f_j\right) \right]\right).$$

2.3   Extensions of Poisson Processes

There are several variants of the Poisson process have studied and we briefly consider some of them.

Compound Poisson Process

The compound Poisson process is used to calculate the total claims on an insurance company. Suppose that a insurance company receives in the time interval $(0, t]$ a random number of claims for according to a simple Poisson process and suppose that the amounts of claims are i.i.d random variables. Then the insurance company is interested in knowing $X(t)$, the total amount of claims it will have to pay in the time interval $[0, t]$. The definition of the process $X(t)$ as follows

Definition 2.4 A stochastic process $\{X_t; t \geq 0\}$ is said to be a *compound Poisson process* if it can be written as

$$X_t = \sum_{i=1}^{N_t} Y_i, \quad t \geq 0, \tag{2.6}$$

where $\{N_t; t \geq 0\}$ is a Poisson process with parameter λ and $\{Y_i; \ i = 1, 2, \dots\}$ are independent and identically distributed random variables. The process $\{N_t; t \geq 0\}$ and the random variables $\{Y_i; \ i = 1, 2, \dots\}$ are assumed to be independent.

Remark 2.7 We have

1. If $Y_t = 1$, then $X_t = N_t$ is a Poisson process.
2. Mean value

$$E(X_t) = E\left(E(X_t|N_t)\right)$$
$$= E\left(E\left(\sum_{i=1}^{N_t} Y_i | N_t\right)\right)$$
$$= E\left(N_t E(Y_i)\right)$$
$$= \lambda t E(Y_i).$$

3. Variance

$$Var(X_t) = E\left(Var(X_t|N_t)\right) + Var\left(E(X_t|N_t)\right)$$
$$= E\left(N_t Var(Y_i)\right) + Var\left(N_t E(Y_i)\right)$$
$$= \lambda t Var(Y_i) + E(Y_i)^2 \lambda t$$
$$= \lambda t \left(Var(Y_i) + E(Y_i)^2\right)$$
$$= \lambda t \left(E(Y_i^2)\right).$$

4. The characteristic function, for any $t \geq 0$,

$$\phi_{X_t}(u) = e^{\lambda t (\phi_Y(u) - 1)}.$$

Remark 2.8 Let $\{Z_t; t \geq 0\}$ be a compound Poisson process and let G be the distribution function of the variables random Y_i. Then if $x \in \mathbb{R}$ we have

$$P\left(Z_t \leq x\right) = \sum_{n=0}^{\infty} P\left(Z_t \leq x \mid N_t = n\right) P\left(N_t = n\right)$$

$$= \sum_{n=0}^{\infty} P\left[\left(\sum_{i=1}^{n} Y_i\right) \leq x\right] P\left(N_t = n\right)$$

$$= \sum_{n=0}^{\infty} G^{*n}\left(x\right) P\left(N_t = n\right)$$

where G^{*n} denotes the nth convolution of G with itself. Calculating the distribution of the random variable Z_t can be complicated since, generally, the convolution G^{*n} does not have a closed form. However, if the common distribution of the random variables Y_i belongs to the family $(a, b, 0)$ of distributions, that is, if for $k \in \mathbb{Z}^+$ we have that

$$P\left(Y_1 = k\right) = \left(a + \frac{b}{k}\right) P\left(Y_1 = k - 1\right)$$

where $a, b \in \mathbb{R}$, then for all $n \in \mathbb{N}$ it is satisfied that:

$$P\left(Z_t = n\right) = \frac{\lambda}{n} \sum_{k=1}^{n} k P\left(Y_1 = k\right) P\left(Z_t = k - 1\right)$$

with

$$P\left(Z_t = 0\right) = \exp\left(-\lambda t\right)$$

The proof of this result can be found in [6].

Mixed Poisson process

The class of Poisson processes referred to as a mixed Poisson process and is used in many applications. We generalize the definition of a Poisson process by assuming that the arrival rate function λ is a random variable (see [7]). The mixed Poisson process has stationary but not independent increments.

Definition 2.5 Let Λ be a positive random variable. Suppose $\{N_t; t \geq 0\}$ is a Poisson process independent of Λ. Then the process $\{N_t; t \geq 0\}$ with rate Λt is called a *mixed Poisson process*.

Let $\{N_t; t \geq 0\}$ be a mixed Poisson process with rate Λ. We assume that the random variable Λ has gamma distributions with shape parameter α and scale parameters β respectively. Then the probability function of N_t is given by [8]:

$$P(N_t = n) = \int_0^\infty e^{-\lambda} \frac{\lambda^n}{n!} \frac{\beta^\alpha}{\Gamma(\alpha)} \lambda^{\alpha-1} e^{-\beta\lambda} d\lambda$$

$$= \binom{\alpha+n-1}{n} \left(\frac{t}{t+\beta}\right)^n \left(\frac{b}{t+\beta}\right)^\alpha$$

$$= NB(\alpha, \beta)$$

We see that $\{N_t; t \geq 0\}$ follows a Pascal or negative binomial distribution with

$$E(N_t) = \frac{\alpha t}{\beta}$$

$$Var(N_t) = \frac{\alpha t}{\beta} + \frac{t^2\alpha}{\beta^2}$$

Suppose if we take $\alpha = 2$ and $\beta = 1$, then the moments are:

$$E(N_t) = 2t$$

$$Var(N_t) = 2t + 2t^2$$

The sample of the simulated mixed Poisson process shown in the Fig. 2.5.

```
     # Set the start and end points of gamma distrbution
a <- 2
b <- 1

# Generate n=100 prior gamma random values

n=100
l<-rgamma(n,a,1/b)

# Generate the process
Nt <- NULL
for(i in 1:n){
   Nt[i]<-rpois(1,i*l[i])
}

# Plot
library(ggplot2)
qplot(1:n,Nt,xlab="Time",ylab="N_t",
      main="Mixed Poisson process",geom=c("step","point"))
```

Fig. 2.5 Mixed Poisson
process

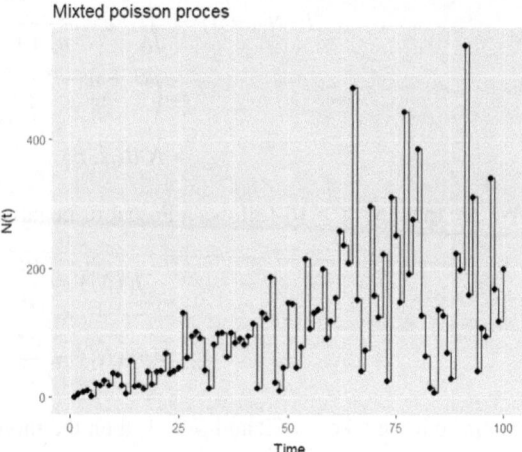

We now develop a method to estimate the intensity function using linear regression (for more details see [9]). We explain the method in the following steps:

1. First, we define a dependent variable, in our case the intensity function, and also a co-variable, in this case, the time t.
2. Define a utility function or a risk function, the most used risk function is the error sum squares [9]. When we have an associate distribution, we can use the likelihood function which is a utility function. The difference between a utility function and a risk function is to find the maximum or minimum argument. We define the utility function

$$R(\alpha, \beta) := \sum_{i=1}^{n}(y_i - (\alpha + \beta x_i))^2$$

then, we have

$$\begin{pmatrix} \hat{\alpha} \\ \hat{\beta} \end{pmatrix} = \underset{\alpha,\beta \in \mathbb{R}}{argmin} \left\{ \sum_{i=1}^{n}(y_i - (\alpha + \beta x_i))^2 \right\}$$

3. We have an explicit expression to solve the minimum argument

$$\hat{\beta} = \frac{\sum_{i=1}^{n}(x_i - \bar{x})(x_i - \bar{y})}{\sum_{i=1}^{n}(x_i - \bar{x})^2}$$

$$\hat{\alpha} = \bar{y} - \hat{\beta}\bar{x}.$$

We now calculate the number of COVID-19 cases in Bogota using a non-homogeneous Poisson process. We consider the reported number of COVID-19 infection data for 12 weeks from August to October 2020 as given in Table 2.1 and the intensity of COVID-19 cases per week (Fig. 2.6).

Table 2.1 Intensity function per week

Week	Intensity
1	2900
2	3160
3	4657
4	8649
5	13102
6	13662
7	20961
8	25014
9	24411
10	29687
11	24954
12	20991

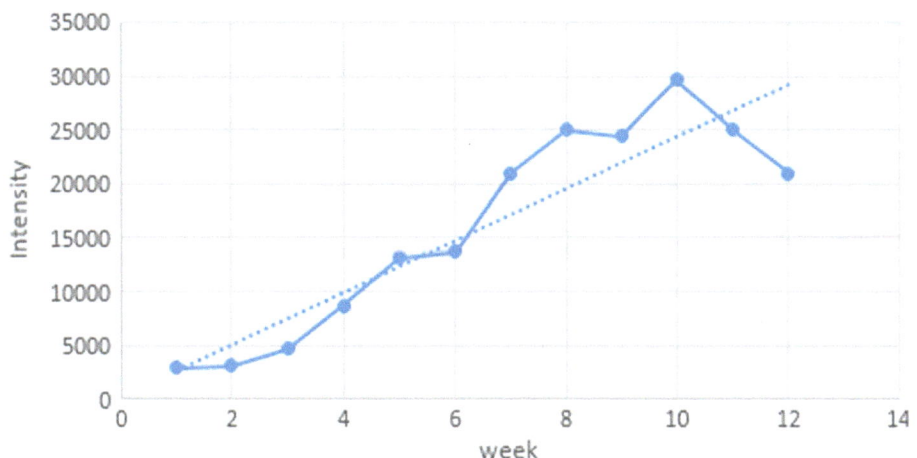

Fig. 2.6 Intensity function of COVID-19 per week

We have obtained the estimated value of the intensity function and is given by

$$\hat{\lambda}(t) = 280.5 + 2420.3t, \, t > 0$$

and the cumulative intensity function is given by

```
      > #LINEAR MODEL
> t<-1:12
> l<-c(2900,3160,4657,8649,13102,13662,20961,25014
+      ,24411,29687,24954,20991)
> fit<-lm(l~t)
> summary(fit)

Call:
lm(formula = l ~ t)

Residuals:
    Min      1Q  Median      3Q     Max
-8332.9 -1952.5  -470.5  2695.6  5371.2

Coefficients:
            Estimate Std. Error t value Pr(>|t|)
(Intercept)    280.5     2495.5   0.112    0.913
t             2420.3      339.1   7.138 3.15e-05 ***
---
Signif. codes:  0 `***' 0.001 `**' 0.01 `*' 0.05 `.' 0.1 ` ' 1

Residual standard error: 4055 on 10 degrees of freedom
Multiple R-squared:   0.8359, Adjusted R-squared:   0.8195
F-statistic: 50.95 on 1 and 10 DF,   p-value: 3.149e-05
```

$$\hat{\Lambda}(t) := \int_0^t \hat{\lambda}(u)du$$

$$= \int_0^t (280.5 + 2420.3u)du, t > 0$$

$$= 280.5t + 1210.15t^2, t > 0$$

$$P(N_t = k) := \frac{\left(\hat{\Lambda}(t)\right)^k}{k!} e^{-\hat{\Lambda}(t)}$$

$$= \frac{\left(280.5t + 1210.15 * t^2\right)^k}{k!} e^{-280.5t + 1210.15t^2}$$

and for $t > 0$

$$\sigma(t) := \sqrt{\Lambda(t)}$$

$$= \sqrt{280.5t + 1210.15t^2}.$$

References

1. Guttorp, P., & Thorarinsdottir, T. L. (2012). What happened to discrete chaos, the Quenouille process, and the sharp Markov property? Some history of stochastic point processes. *International Statistical Review, 80*(2), 253–268.
2. Penttinen, A., Stoyan, D., & Henttonen, H. M. (1992). Marked point processes in forest statistics. *Forest Science, 38*(4), 806–824.
3. Ross, S. M. (1996). *Stochastic processes* (2nd ed.). New York: Wiley.
4. Webel, K., & Wied, D. (2016). *Stochastische Prozesse: Eine Einführung für Statistiker und Datenwissenschaftler* (segunda ed.). Wiesbaden: Springer.
5. Drazek, L. C. M. T. (2013). Intensity estimation for Poisson processes. School of Mathematics, University of Leeds, Leeds.
6. Dickson, D. C. (2005). *Insurance risk and ruin* (2nd ed.). New York: Cambridge University Press.
7. Resnick, S. I. (2001). *Adventures in stochastic processes* (2nd ed.). Boston: Birkhäuser.
8. Teugels, J. L., & Vynckier, P. (1996). The structure distribution in a mixed Poisson process. *International Journal of Stochastic Analysis, 9*(4), 1–8. https://doi.org/10.1155/S1048953396000421.
9. Munandar, D., Supian, S., & Subiyanto, S. (2020). Probability distributions of COVID-19 tweet posted trends uses a nonhomogeneous Poisson process. *International Journal of Quantitative Research and Modeling, 1*(4), 229–238.

Continuous-Time Markov Chain Modeling

As we mentioned earlier, the Russian mathematician Andrei Andreyevich Markov (1856–1922) introduced sequences of values of a random variable in which the value of the variable in the future depends on the value of the variable in the present but is independent of the history of the past. These sequences are known today as Markov chains. Markov chains with discrete-time parameters are frequently used models in sociology, medicine, and biology because the simulation is easier to build in discrete steps. However, in many situations, time runs continuously. The discrete model may not be the most appropriate, for example, in the description of the development of infectious-contagious diseases, in which the number of susceptible (S), infected (I), and removable (R) individuals depends on the pattern of contacts between members of the population. Consequently, the corresponding population sizes in each compartment S, I, and R vary continuously over time. In such situations, Markov chains with a continuous time parameter turn out to be more suitable models.

The theory of the continuous-time Markov chain is similar to the discrete-time Markov chain. In this chapter, we will present the basic concepts of the continuous-time Markov chain and some applications. We introduce Birth and death processes, a class of continuous-time Markov chain used to model population biology.

3.1 Introduction: Definition and Basic Properties

Definition 3.1 Let $\{X_t : t \in [0, \infty)\}$ be a stochastic process with countable set of state space $S \subseteq \mathbb{N}$ and $S \neq \emptyset$. We say that $\{X_t : t \in [0, \infty)\}$ is a continuous-time Markov chain if it satisfies the following property known as the *Markov property:* For any finite or countable subset of points $0 \leq t_0 < t_1 < t_2 < \cdots < t_n < t$ with $t_0, t_1, \ldots, t_n, t \in [0, \infty)$ and for all $i_o, i_1, i_{n-1}, i, j \in S$ we have that

© The Author(s), under exclusive license to Springer Nature Switzerland AG 2023
L. Blanco-Castañeda and V. Arunachalam, *Applied Stochastic Modeling*, Synthesis Lectures on Mathematics & Statistics, https://doi.org/10.1007/978-3-031-31282-3_3

$$P(X_t = j \mid X_{t_n} = i, X_{t_{n-1}} = i_{n-1}, \ldots, X_{t_0} = i_0.) = P\left(X_t = j \mid X_{t_n} = i\right). \qquad (3.1)$$

Definition 3.2 We define the following probability

$$p_{ij}(t) := P(X_{t+s} = j \mid X_s = i) \qquad (3.2)$$

is the transition probability from a state $i \in S$ to another state $j \in S$ after a duration time t.

The transition probability satisfies the following conditions

a. $p_{ij}(0) = \delta_{ij} := \begin{cases} 1 \text{ if } i = j \\ 0 \text{ if } i \neq j \end{cases}$

b. $\lim\limits_{t \to 0^+} p_{ij}(t) = \delta_{ij}$

c. For any $t \geq 0$, $i, j \in S$ we have $0 \leq p_{ij}(t) \leq 1$ and $\sum\limits_{k \in S} p_{ik}(t) = 1$.

d. For all $i, j \in S$ and and $s, t \geq 0$ we have that

$$p_{ij}(t + s) = \sum_{k \in S} p_{ik}(t) \cdot p_{kj}(s)$$

We consider only homogeneous Markov chains with transition probabilities, for which the probabilities given in the above definition do not depend on s for $s < t$. The transition probability matrix is denoted by $P(t) :- (p_{ij}(t))_{i,j \in S}, t \in T$, a stochastic matrix.

Remark 3.1 Analogously as in the discrete case, it can be seen that in this case the Chapman-Kolmogorov equation, is given by

$$p_{ij}(t + s) = \sum_{k \in S} p_{ik}(t) p_{kj}(s) \qquad (3.3)$$

for all $i, j \in S$ and for all $s, t \geq 0$. In the matrix form

$$P(t + s) = P(t)P(s)$$

That is, the family of transition matrices forms a semi-group.

Suppose that $\{X_t; t \geq 0\}$ is a Markov chain with state space S and let arbitrary $i \in S$ be fixed. Let T_i the length of time the chain in state i before transitioning to another state. We have

$$P\left(T_i > s + t \mid T_i > s\right) = P\left(X_u = i,\ 0 \le u \le s + t \mid X_u = i,\ 0 \le u \le s\right)$$
$$= \frac{P\left(X_u = i,\ 0 \le u \le s + t,\ X_u = i,\ 0 \le u \le s\right)}{P\left(X_u = i,\ 0 \le u \le s\right)}$$
$$= \frac{P\left(X_u = i,\ s \le u \le s + t,\ X_u = i,\ 0 \le u \le s\right)}{P\left(X_u = i,\ 0 \le u \le s\right)}$$
$$= P\left(X_u = i,\ s \le u \le s + t \mid X_u = i,\ 0 \le u \le s\right)$$
$$= P\left(X_u = i,\ s \le u \le s + t \mid X_s = i\right)$$
$$= P\left(T_i > t\right).$$

This implies that the random variables T_i has an exponential distribution.

We now state the following results without proof, and the interested reader may refer to [1].

1. $p_{ij}(t)$ is uniformly continuous on $[0, \infty)$.
2. For each $i \in S$ we have that

$$\lim_{t \to 0^+} \frac{1 - p_{ii}(t)}{t} = q_i \tag{3.4}$$

   exists (but may be equal to $+\infty$).
3. For all $i, j \in S$ with $i \ne j$, we have that the following limit exists:

$$\lim_{t \to 0^+} \frac{p_{ij}(t)}{t} = q_{ij} < \infty. \tag{3.5}$$

4. The matrix $Q = \left(q_{ij}\right)_{i,j \in S}$ is called the infinitesimal generator of the Markov chain and written as;

$$Q = \begin{bmatrix} -q_0 & q_{01} & q_{02} & \cdots \\ q_{10} & -q_1 & q_{12} & \cdots \\ q_{20} & q_{21} & -q_2 & \cdots \\ \vdots & \vdots & \vdots & \end{bmatrix}$$

   with

$$Q = P'(0).$$

We wish to state that the probability of transition from the state i to state j, with $i \ne j$, on an interval of the small interval of time Δt, is approximately equal to $q_{ij}\Delta t$; while the probability that the chain remains in state i is approximately equal to $q_{ii}\Delta t + 1$. The values q_{ij} with $i \ne j$ and q_{ii} are called, respectively, transition rates and retention rates.

Lemma 3.1 *Let $\{X_t; t \ge 0\}$ be a continuous-time Markov chain with state states S, the transition matrix $P(t)$, $t \ge 0$ and the infinitesimal generator Q satisfies the following:*

a. $0 \leq q_{ij} < \infty$, $i \neq j$; $-\infty \leq q_{ii} \leq 0$

b. $\sum_{j \neq i} q_{ij} \leq -q_{ii} =: q_i$

If the state space S is finite, then the transition matrix satisfies for all $i \in S$:

$$q_{ii} = \sum_{j \neq i} q_{ij}$$

Proof a. $0 \leq p_{ij}(t) \leq 1$ then we have that $0 \leq q_{ij}$ for all $i \neq j$ and $q_{ii} \leq 0$.
Also $q_{ij} < \infty$ (see [1]).

b. For a fixed $t \geq 0$ and $i \in S$, we know that

$$\sum_{j \in S} p_{ij}(t) = 1$$

If $t > 0$ we have

$$\frac{1}{t} \sum_{i \neq j, j \in S} p_{ij}(t) = \frac{1 - p_{ii}(t)}{t}$$

After some calculations using the Fatou lemma, we get

$$-q_{ii} = \lim_{t \to 0^{!}} \frac{1 - p_{ii}(t)}{t}$$

$$= \lim_{t \to 0^+} \frac{1}{t} \sum_{i \neq j, j \in S} p_{ij}(t)$$

$$\geq \sum_{i \neq j, j \in S} \limsup_{t \to 0^+} \frac{p_{ij}(t)}{t}$$

$$= \sum_{i \neq j, j \in S} q_{ij}.$$

Remark 3.2 The *state $i \in S$ is regular or stable*, if $q_{ii} < \infty$ and $\sum_{j \neq i} q_{ij} = q_{ii}$. Otherwise, it is called *non-regular*. The state $i \in S$ is *instantaneous*, if $q_{ii} = \infty$ and absorbing if $q_{ii} = 0$.

We now discuss Kolmogorov differential equations for the continuous-time Markov chain, which plays an essential role in modeling and is a necessary part of understanding the next section on the Birth and Death process. We now state the following theorem.

Theorem 3.1 *Suppose that $q_{ii} < \infty$ for each $i \in S$, then the transition probabilities $p_{ij}(t)$ are differentiable for all $t \geq 0$ and for each $i, j \in S$ and satisfies the following respectively, Kolmogorov backward and forward equations given by*

$$\frac{dp_{ij}(t)}{dt} = -q_{ii}p_{ij}(t) + \sum_{\substack{k \in S \\ k \neq i}} q_{ik}p_{kj}(t) \tag{3.6}$$

and

$$\frac{dp_{ij}(t)}{dt} = -p_{ij}q_{jj}(t) + \sum_{\substack{k \in S \\ k \neq i}} p_{ik}q_{kj}(t) \tag{3.7}$$

Proof Consider a Markov chain with finite state space S. Using the Chapman-Kolmogorov equation, for all $i, j \in S$ and for all $t, h > 0$ we have

$$p_{ij}(t+h) = \sum_{k \in S} p_{ik}(h) p_{kj}(t)$$

$$p_{ij}(t+h) - p_{ij}(t) = p_{ii}(h) p_{ij}(t) + \sum_{\substack{k \in S \\ k \neq i}} p_{ik}(h) p_{kj}(t) - p_{ij}(t)$$

and now

$$\frac{p_{ij}(t+h) - p_{ij}(t)}{h} = p_{ij}(t)\left(\frac{p_{ii}(h) - 1}{h}\right) + \frac{1}{h}\sum_{\substack{k \in S \\ k \neq i}} p_{ik}(h) p_{kj}(t)$$

taking limit $h \to 0^+$ we get:

$$\frac{dp_{ij}(t)}{dt} = -q_{ii} + \sum_{\substack{k \in S \\ k \neq i}} q_{ik}p_{kj}(t).$$

Similarly, one can prove the Kolmogorov forward equation.

The matrix form of the above differential equations can be written in the matrix form with the initial condition $P(0) = I$:

$$\frac{dP(t)}{dt} = QP(t)$$

and

$$\frac{dP(t)}{dt} = P(t)Q.$$

We now briefly discuss the solution approach for the above system of differential equations. When S is a finite set, the solution of the above equation is given by:

$$P(t) = \exp(Qt) = \sum_{k=0}^{\infty} \frac{(tQ)^k}{k!} \tag{3.8}$$

It can be shown that the above solution is valid provided that the q_i are bounded. Consider a finite-dimensional chain with matrix Q and diagonalizable. Assume that the eigenvalues $\beta_0, \beta_1, \ldots, \beta_n$ are all distinct of the matrix Q. Then there exists a nonsingular matrix H such that Q can be written in the form

$$Q = H \begin{pmatrix} \beta_0 & & 0 \\ & \ddots & \\ 0 & & \beta_n \end{pmatrix} H^{-1}$$

and the solution matrix:

$$P(t) = H \begin{pmatrix} e^{\beta_0 t} & & 0 \\ & \ddots & \\ 0 & & e^{\beta_n t} \end{pmatrix} .H^{-1} .$$

We wish to note that in general, the eigenvalues of the matrix Q are not necessarily distinct, but still, Q can be expressed in the form given in (3.8), and the $P(t)$ can be obtained as above. The analytical solution can be calculated when the system of equations is small. For a large system of equations, one can obtain the transient solution, or by numerical methods using programs Python, or R.

Example 3.1 A system can be found in any of the two states namely a "free" state denoted by (0) and "a busy" state denoted by (1). Suppose that the time spends in each of the states is random variables with exponential distributions with parameters λ and μ respectively. In that case, we have

$$q_{00} = -\lambda, \quad q_{01} = \lambda$$
$$q_{10} = \mu, \quad q_{11} = -\mu$$

And, the infinitesimal generator is given by:

$$Q = \begin{pmatrix} -\lambda & \lambda \\ \mu & -\mu \end{pmatrix}$$

Therefore, with backward Kolmogorov equations, we obtain that:

$$\frac{dp_{00}(t)}{dt} = -\lambda_{00}(t) + \lambda_{10}(t)$$

$$\frac{dp_{01}(t)}{dt} = -\lambda_{01}(t) + \lambda_{11}(t,)$$

$$\frac{dp_{10}(t)}{dt} = -\mu_{10}(t) + \mu_{00}(t)$$

$$\frac{dp_{11}(t)}{dt} = -\mu_{11}(t) + \mu_{01}(t)$$

since:

$$p_{01}(t) = 1 - p_{00}(t)$$

and

$$p_{11}(t) = 1 - p_{10}(t).$$

Using the matrix notation, we have

$$\frac{dY(t)}{dt} = QY(t)$$

where,

$$Y(t) = \begin{pmatrix} p_{00}(t) \\ p_{10}(t) \end{pmatrix}$$

with the initial condition,

$$Y(0) = \begin{pmatrix} 1 \\ 0 \end{pmatrix}$$

Since Q's eigenvalues are 0 and $-(\lambda + \mu)$ and Q's eigenvectors

$$v = \begin{pmatrix} 1 \\ 1 \end{pmatrix} \text{ and } \omega = \begin{pmatrix} -\lambda \\ \mu \end{pmatrix}$$

we get the solution of the system and given by

$$\begin{pmatrix} p_{00}(t) \\ p_{10}(t) \end{pmatrix} = \alpha \begin{pmatrix} 1 \\ 1 \end{pmatrix} + \beta \begin{pmatrix} -\lambda \\ \mu \end{pmatrix} \exp\left(-(\lambda + \mu)t\right)$$

since

$$\begin{pmatrix} p_{00}(0) \\ p_{10}(0) \end{pmatrix} = \begin{pmatrix} 1 \\ 0 \end{pmatrix}$$

we can obtain:

$$p_{00}(t) = \frac{\mu}{\lambda + \mu} + \frac{\lambda}{\lambda + \mu} \exp\left(-(\lambda + \mu)t\right)$$

$$p_{10}(t) = \frac{\mu}{\lambda + \mu} - \frac{\mu}{\lambda + \mu} \exp\left(-(\lambda + \mu)t\right)$$

Note that if $t \to \infty$ then

$$\lim_{t \to \infty} P(t) = \begin{pmatrix} \frac{\mu}{\lambda + \mu} & \frac{\lambda}{\lambda + \mu} \\ \frac{\mu}{\lambda + \mu} & \frac{\lambda}{\lambda + \mu} \end{pmatrix}$$

The following question naturally arises: given a Markov chain with continuous-time parameter, we are interested to know the conditions for the existence of limit $\lim_{t \to \infty} p_{ij}(t)$. For this, we introduce the following notation and concepts:

Let $\{X_t; t \geq 0\}$ a Markov chain with discrete state space S and consider

$$T_0 := 0$$
$$T_1 := \inf \{t \geq 0 : X_t \neq X_0\}$$
$$T_2 := \inf \{t > T_1 : X_t \neq X_{T_1}\}$$
$$\vdots$$
$$T_n := \inf \{t > T_{n-1} : X_t \neq X_{T_{n-1}}\}$$

where, T_1 represents the time at which the chain first changes its initial state, T_2 is when the chain changes its state for the second time, and so on. For $n \geq 1$ we consider the random variables

$$\tau_n := T_n - T_{n-1}$$

and we define the immersed Markov chain $Y = (Y_n)_{n \in \mathbb{N}}$ as follows:

$$Y_n := X_{\tau_n}$$

we have the following useful theorem stated without proof.

Theorem 3.2 *Let $X = \{X_t; t \geq 0\}$ a Markov chain with a states set S and infinitesimal generator $Q = (q_{ij})_{i,j \in S}$ such that*

1. $q_{ij} \geq 0$ *for all $i, j \in S$ with $i \neq i$ $q_{ii} < 0$ for all $i \in S$.*
2. $\sum_{j \in S} q_{ij} = 0$ *for all $i \in S$*
3. $0 < \sup_{i \in S} |q_{ii}| = \sup_{i \in S} q_i < \infty$

then, it satisfies that $Y = (Y_n)_{n \in \mathbb{N}}$ is a Markov chain on a discrete parameter space S and transition matrix $\mathbf{P} = (p_{ij})_{i,j \in S}$ where

$$p_{ij} = \begin{cases} \frac{q_{ij}}{q_i} & \text{if } i \neq j \\ 0 & \text{if } i = j \end{cases}, q_i > 0$$

and

$$p_{ij} = \delta_{ij} , \ q_i = 0$$

Example 3.2 Let $\{X_t; t \geq 0\}$ a Markov chain on a continuous space with a state set given by $S = \{a, b, c\}$ and the infinitesimal generator is given by

$$Q = \begin{pmatrix} -5 & 3 & 2 \\ 1 & -2 & 1 \\ 4 & 0 & -4 \end{pmatrix}$$

then the chain $Y = (Y_n)_{n \in \mathbb{N}}$ has a states set given by S and transition matrix given by:

$$P = \begin{pmatrix} 0 & \frac{3}{5} & \frac{2}{5} \\ \frac{1}{2} & 0 & \frac{1}{2} \\ 1 & 0 & 0 \end{pmatrix}.$$

Remark 3.3 A Markov chain in continuous time, the infinitesimal generator Q satisfies the conditions 1, 2 and 3 of the Theorem 3.2 is called a regular Markov chain. Condition (3.) ensures that infinitely many state changes do not occur in the interval of finite length; in other words, it is required that the number of transitions that the chain makes in an interval of finite length is a finite amount.

We can observe it $q_i > 0$ then $Q = (q_{ij})_{i,j \in S}$ has the components

$$q_{ij} = \begin{cases} q_i \, p_{ij} & \text{if } i \neq j \\ -q_i \, (1 - p_{ii}) & \text{if } i = j \end{cases}$$

for each states $i, j \in S$.

Definition 3.3 Let $\{X_t; t \geq 0\}$ a Markov chain with a discrete states space S and if $i \in S$ with $q_i < \infty$. We say that

1. i is recurrent if i is recurrent for the immersed chain $Y = (Y_n)_{n \in \mathbb{N}}$
2. i is positive recurrent if i is recurrent and the expected time m_i of the first return to state i is finite.
3. $\{X_t; t \geq 0\}$ is irreducible if immersed chain $\{Y_n; n \in \mathbb{N}\}$ is irreducible.

Definition 3.4 Let $\{X_t; t \geq 0\}$ a Markov chain with a discrete states set S and transition matrix $(P(t))_{t \geq 0}$. A measure $\mu : S \longrightarrow [0, \infty)$ is called *invariant*, is and only if, for all $t \geq 0$ and for every $j \in S$ it satisfies then:

$$\mu_j = \sum_{i \in S} \mu_i \, p_{ij}(t). \tag{3.9}$$

with $\sum_{i \in S} \mu_i = 1$. We call μ the stationary distribution of the Markov chain on S.

In matrix notation, we write

$$\mu = \mu P(t)$$

We call μ the stationary distribution of the Markov chain on S.

Theorem 3.3 *Let* $X := \{X_t; t \geq 0\}$ *a regular Markov chain with a discrete states space* S *and infinitesimal generator* $Q = (q_{ij})_{i,j \in S}$. *Then* μ *is an invariant measure on* S *if and only if:*

$$\mu Q = 0$$

Proof Consider

$$\frac{dP(t)}{dt} = P(t)Q$$

and

$$\frac{dP(t)}{dt} = QP(t)$$

Then

$$P(t) = P(0) + \int_0^t \frac{dP(s)}{ds} ds$$

$$= P(0) + \int_0^t P(s)Qds$$

$$= P(0) + \left(\int_0^t P(s) ds \right) Q \tag{3.10}$$

and

$$P(t) = P(0) + Q\left(\int_0^t P(s) ds \right). \tag{3.11}$$

If μ is an invariant measure for $(X_t)_{t\geq 0}$ then for all $t \geq 0$ it satisfies

$$\mu = \mu P(t)$$

therefore (3.10) we have for all $t \geq 0$:

$$\mu = \mu P(0) + t\mu Q$$

$$= \mu + t\mu Q$$

hence,

$$\mu Q = 0$$

reciprocally if $\mu Q = 0$ then (3.11) we can obtain

$$\mu P(t) = \mu P(0) + \mu Q\left(\int_0^t P(s) ds \right)$$

$$= \mu$$

We conclude that μ is an invariant measure for $(X_t)_{t\geq 0}$.   □

Theorem 3.4 *Let* $X := \{X_t; t \geq 0\}$ *a regular irreducible Markov chain with matrix* Q. *Suppose that* $v = (v_i)_{i\in S}$ *an invariant measure for the immersed chain* $(Y_n)_{n\in\mathbb{N}}$, *then* $\mu = (\mu_i)_{i\in S}$ *with* $\mu_i = \frac{v_i}{q_i}$ *is an invariant measure for* $X = \{X_t; t \geq 0\}$. *Reciprocally, if* μ *is an invariant measure for* $X = \{X_t; t \geq 0\}$ *then* $v = (q_i\mu_i)_{i\in S}$ *is an invariant measure for* $(Y_n)_{n\in\mathbb{N}}$.

Proof Using the hypothesis all states are regular. Let L the diagonal matrix with components q_i with $i \in S$. As earlier, we constructed the immerse Markov chain, we have

$$Q = L\,(\mathbf{P} - I)$$

where $\mathbf{P}$ is the transition matrix of the immerse Markov chain $(Y_n)_{n\in\mathbb{N}}$ and I is the identity matrix.

Using the definition of the invariant measure, we have μ for $X = \{X_t; t \geq 0\}$ if and only if

$$0 = \mu Q = \mu L\,(\mathbf{P} - I)$$
$$\mu L = \mu L \mathbf{P}$$

That is, $\mu L = (q_i \mu_i)_{i \in S}$ is an invariant measure for $(Y_n)_{n\in\mathbb{N}}$. $\qquad\square$

Definition 3.5 Let $\{X_t; t \geq 0\}$ be a continuous-time Markov chain. We say that the chain is reversible, if and only if, the immerse Markov chain is also reversible.

Remark 3.4 Using the above definition we have, $\{X_t; t \geq 0\}$ is reversible, if and only if, for all $i, j \in S$ it satisfies that

$$\rho_i\, p_{ij} = \rho_j\, p_{ji}$$

where $\rho = (\rho_i)_{i\in S}$ a stationary distribution of the immerse Markov chain and $(p_{ij})_{i,j\in S}$ the transition matrix of the immerse Markov chain. Therefore, $(X_t)_{t\geq 0}$ is reversible, if and only if, for all $i, j \in S$ we have:

$$\rho_i\frac{q_{ij}}{q_i} = \rho_j\frac{q_{ji}}{q_j}$$
$$v_i q_{ij} = v_j q_{ji}$$

where $v_i := \frac{\rho_i}{q_i}$ with $i \in S$. Using the theory developed earlier, we have $(v_i)_{i\in S}$ is an invariant measure of the Markov chain $\{X_t; t \geq 0\}$.

Example 3.3 The birth and death process is a reversible process. See the Example 1.18 presented in Chap. 1.

Remark 3.5 Using the previous theorem we have $P = (p_i)_{i\in S}$ where

$$p_i = \frac{\mu_i}{\displaystyle\sum_{j\in S}\mu_j} = \frac{\frac{v_i}{q_i}}{\displaystyle\sum_{j\in S}\frac{v_j}{q_j}}$$

is a stationary distribution for the irreducible regular Markov chain $X = \{X_t; t \geq 0\}$. If $\{X_t; t \geq 0\}$ is a regular Markov chain, positive, recurrent and irreducible, then

$$\lim_{t \to \infty} P(X_t = j) = p_j$$

exist for all $j \in S$ and it is independent of the initial distribution.

For the stationary distribution of the Markov chain, we give the following theorem without proof.

Theorem 3.5 *Let $X = \{X_t; t \geq 0\}$ a regular Markov chain with matrix Q. If the chain is irreducible and all states are recurrent, then:*

a.

$$\lim_{t \to \infty} p_{ij}(t) = \frac{1}{m_j q_j}$$

for all $i, j \in S$ regardless of initial state $i \in S$ and where m_j is the expected time of the first return of the chain to state j,

b. if the chain has a positive recurrent state j, then exist a unique stationary distribution $P = (p_i)_{i \in S}$ on S. In that case we have

$$p_i = \frac{1}{m_i q_i}$$

for all $i \in S$ and all recurrent positive states.

3.2   Birth and Death Processes

A *birth and death process* with state space $S = \mathbb{N}$ is an important tool for modeling queueing, reliability and population biology. Consider a continuous-time Markov chain $X = (X_t)_{t \geq 0}$ with states $S = \mathbb{N}$ and infinitesimal generator matrix

$$Q = (q_{ij})_{i,j \in S}$$

with

$$q_{ij} = \begin{cases} \lambda_i & \text{if } j = i + 1 \\ \mu_i & \text{if } j = i - 1 \\ -(\lambda_i + \mu_i) & \text{if } i = j \\ 0 & \text{in other case} \end{cases}$$

Where λ_i and μ_i are non-negative reals numbers, respectively, birth and death rates.

If all $\mu_i = 0$, the process is called pure birth process, on the other hand, if all $\lambda_i = 0$, the process is called pure death process. A pure birth process with constant birth rates $\lambda_i = \lambda$ for all i, is a Poisson process with parameter λ.

The infinitesimal generator in a birth and death process is given by:

$$Q = \begin{pmatrix} -\lambda_0 & \lambda_0 & 0 & 0 & \cdots \\ \mu_1 & -(\lambda_1 + \mu_1) & \lambda_1 & 0 & \cdots \\ 0 & \mu_2 & -(\lambda_2 + \mu_2) & \lambda_2 & \cdots \\ 0 & 0 & \mu_3 & -(\lambda_3 + \mu_3) & \cdots \\ \vdots & \vdots & \vdots & \vdots & \vdots \end{pmatrix}$$

The transition probabilities of the immerse Markov chain $(Y_n)_{n \in \mathbb{N}}$ are given by:

$$p_{ij} = \begin{cases} \frac{\lambda_i}{\lambda_i + \mu_i} & \text{if} \quad j = i + 1 \\ \frac{\mu_i}{\lambda_i + \mu_i} & \text{if} \quad j = i - 1 \\ 1 & \text{if } i = 0, \ j = 1 \\ 0 & \text{in} \quad \text{other case} \end{cases}$$

With the goal to finding the stationary distribution, if it exists, we need to solve the equation $PQ = 0$, in other words:

$$0 = -\lambda_0 p_0 + \mu_1 p_1$$
$$0 = \lambda_0 p_0 - (\mu_1 + \lambda_1) p_1 + \mu_2 p_2$$
$$\vdots$$
$$0 = \lambda_i p_i - (\mu_{i+1} + \lambda_{i+1}) p_{i+1} + \mu_{i+2} p_{i+2}, \ i \geq 1$$

where we can obtain:

$$p_1 = \frac{\lambda_0}{\mu_1} p_0$$
$$p_2 = \frac{1}{\mu_2} \left(\frac{(\mu_1 + \lambda_1) \lambda_0}{\mu_1} p_0 - \frac{\lambda_0 \mu_1}{\mu_1} p_0 \right) = \frac{\lambda_0 \lambda_1}{\mu_1 \mu_2} p_0$$
$$\vdots$$

This is,

$$p_{i+1} = \frac{\lambda_0 \lambda_1 \cdots \lambda_i}{\mu_1 \mu_2 \cdots \mu_{i+1}} p_0, \ i \geq 0$$

we have

$$\sum_{i \in S} p_i = 1,$$

if and only if,

$$1 = p_0 + p_0 \sum_{i=0}^{\infty} \frac{\lambda_0 \lambda_1 \cdots \lambda_i}{\mu_1 \mu_2 \cdots \mu_{i+1}}$$

if and only if,

$$1 = p_0 \left(1 + \sum_{i=0}^{\infty} \frac{\lambda_0 \lambda_1 \cdots \lambda_i}{\mu_1 \mu_2 \cdots \mu_{i+1}} \right)$$

if and only if,

$$p_0 = \left(1 + \sum_{i=0}^{\infty} \frac{\lambda_0 \lambda_1 \cdots \lambda_i}{\mu_1 \mu_2 \cdots \mu_{i+1}} \right)^{-1}$$

The chain $\{X_t; t \geq 0\}$ is positive recurrent, if and only if, the immerse Markov chain $(Y_n)_{n \in \mathbb{N}}$ is positive recurrent too. According to theory in the discrete Markov chain we have the immersed Markov chain is positive recurrent, if and only if,

$$\sum_{j=1}^{\infty} \prod_{k=0}^{j-1} \left(\frac{p_{k,k+1}}{p_{k+1,k}} \right) < \infty$$

if and only if,

$$\sum_{j=1}^{\infty} \prod_{l=1}^{j} \left(\frac{p_{l-1,l}}{p_{l,l-1}} \right) < \infty$$

$$\sum_{j=1}^{\infty} \prod_{l=1}^{j} \left(\frac{\frac{\lambda_{l-1}}{\lambda_{l-1}+\mu_{l-1}}}{\frac{\mu_l}{\lambda_l+\mu_l}} \right) < \infty$$

$$\sum_{j=1}^{\infty} \prod_{l=1}^{j} \left(\frac{\lambda_{l-1} (\lambda_l + \mu_l)}{\mu_l (\lambda_{l-1} + \mu_{l-1})} \right) < \infty$$

if and only if,

$$\sum_{j=1}^{\infty} \frac{\lambda_0 \lambda_1 \cdots \lambda_j}{\mu_1 \mu_2 \cdots \mu_{j+1}} < \infty$$

the stationary distribution is given by:

$$p_{j+1} = \frac{\lambda_0 \lambda_1 \cdots \lambda_j}{\mu_1 \mu_2 \cdots \mu_{j+1}} p_0, \quad j \in \mathbb{N}$$

The birth and death process with constant rates

Let $\{X_t; t \geq 0\}$ a birth and death process the with $\lambda_i = \lambda > 0$ for all $i \in S$ and $\mu_0 = 0$, $\mu_i = \mu > 0$ for all $i \geq 1$. The process turns out to be positive recurrent, if and only if,

$$\sum_{j=1}^{\infty} \frac{\lambda_0 \lambda_1 \cdots \lambda_j}{\mu_1 \mu_2 \cdots \mu_{j+1}} = \sum_{j=1}^{\infty} \frac{\lambda^{j+1}}{\mu^{j+1}}$$

$$= \sum_{j=1}^{\infty} \left(\frac{\lambda}{\mu}\right)^{j+1} < \infty$$

if and only if,

$$\lambda < \mu$$

The stationary distribution is given by:

$$p_j = \left(\frac{\lambda}{\mu}\right)^j p_0, \quad j = 0, 1, 2, \ldots$$

since

$$1 = \sum_{j=0}^{\infty} p_j$$

$$= p_0 \sum_{j=0}^{\infty} \left(\frac{\lambda}{\mu}\right)^j$$

$$= p_0 \frac{1}{1 - \frac{\lambda}{\mu}}$$

then

$$p_0 = 1 - \frac{\lambda}{\mu}$$

and hence

$$p_j = \left(\frac{\lambda}{\mu}\right)^j \left(1 - \frac{\lambda}{\mu}\right), \quad j \in \mathbb{N}$$

the above limit distribution is geometric with parameter $\frac{\lambda}{\mu}$.

Example 3.4 Let $\{X_t; t \geq 0\}$ a birth and death process with $\lambda_i = \lambda$ and $\mu_i = i\mu$ for all $i \in S$. The resulting process is positive recurrent, if and only if,

$$\sum_{j=1}^{\infty} \frac{\lambda_0 \lambda_1 \cdots \lambda_j}{\mu_1 \mu_2 \cdots \mu_{j+1}} = \sum_{j=1}^{\infty} \frac{\lambda^{j+1}}{(j+1)! \mu^{j+1}}$$

$$= \sum_{j=1}^{\infty} \frac{1}{(j+1)!} \left(\frac{\lambda}{\mu} \right)^{j+1} < \infty$$

In this case, the limit distribution is,

$$p_j = \frac{\lambda_0 \lambda_1 \cdots \lambda_{j-1}}{\mu_1 \mu_2 \cdots \mu_j} p_0$$

$$= \frac{\lambda^j}{j! \mu^j} p_0$$

by way of

$$1 = \sum_{j=0}^{\infty} \frac{\lambda^j}{j! \mu^j} p_0$$

$$1 = p_0 \exp \left(\frac{\lambda}{\mu} \right)$$

the limit distribution is given by:

$$p_j = \frac{1}{j!} \left(\frac{\lambda}{\mu} \right)^i e^{-\frac{\lambda}{\mu}}, \ j \in \mathbb{N}$$

that is to say, the limit distribution is a Poisson with parameter $\frac{\lambda}{\mu}$.

Birth and death process with immigration

Let $\{X_t; t \geq 0\}$ a birth and death process with $\lambda_i = i\lambda + \alpha$ and $\mu_i = i\mu$ for all $i \in S$, with $\alpha > 0$ constant (immigration rate). In this case, the process is positively recurrent if:

$$\sum_{j=1}^{\infty} \frac{\lambda_0 \lambda_1 \cdots \lambda_j}{\mu_1 \mu_2 \cdots \mu_{j+1}} = \sum_{j=1}^{\infty} \frac{\alpha \, (\alpha + \lambda) \, (\alpha + 2\lambda) \cdots (\alpha + j\lambda)}{(j+1)! \mu^j} < \infty$$

in that case, the limit distribution is given by:

$$p_j = \frac{p_0}{j!\mu^j} \prod_{k=0}^{j-1} (\alpha + k\lambda)$$

$$p_0 = \left[1 + \sum_{j=1}^{\infty} \left(\frac{1}{j!\mu^j} \prod_{k=0}^{j-1} (\alpha + k\lambda) \right) \right]^{-1}$$

Let $X_t :=$ "customer number in a system at the time t". It is clear $\{X_t; t \geq 0\}$ is a birth and death process with birth and death rates given by λ_i and $\mu_i, i \in \mathbb{N}$, respectively. If $\lambda_i = \lambda$ and $\mu_i = \mu$ for all i, it known as M/M/1 *queueing models* [2]. If $\lambda_i = \lambda$ for all $i > 0$ and

$$\mu_i = \begin{cases} i\mu \text{ if } 1 \leq i \leq s \\ s\mu \text{ if } \quad i > s \end{cases}$$

with $0 \leq s \leq \infty$, then we will say the M/M/s queueing model.

If $\mu_i = \mu$ and

$$\lambda_i = \begin{cases} (N - i)\lambda \text{ if } i \leq N \\ 0 \quad \text{ if } i > N \end{cases}$$

This corresponds to M/M/1/N queueing model. The M/M/1/N model describes when the population size is infinite and there is only one server. The arrival times to the system and the serviced times are random variables with exponential distribution with parameters λ and μ, respectively. The M/M/1/N model behaves in the same path as the previous system, the only difference being that this allows a maximum customer number finite N in the system. The M/M/s model describes a queue system similar to the M/M/1 system, with the difference being that this model has s servers instead of a unique server.

Example 3.5 There are N children and one nursemaid to take care of them in daycare. From time to time, a baby starts crying, needing the nursemaid's attention, and the probability of such an event in the small time interval $(t, t + h)$ is $\lambda h + o(h)$. If the nursemaid is already busy with a crying baby and another one starts crying, then the crying baby/babies must wait for the nursemaid's attention until she becomes free. The probability that a crying baby who is being attended to will stop crying in the interval $(t, t + h)$ is $\mu h + o(h)$. Let X(t)= number of babies waiting to be attended by nursemaid at time t. Assuming X(0)=0, what is the probability that eventually all the children in the daycare will be calm?

In this case, we have a birth and death process rates $\lambda_i = (N - i)\lambda$ if $0 \leq i \leq N$ and $\mu_i = \mu$ for all $i \geq 1$, and follows M/M/1/N queueing model.

The Kolmogorov backward equations are given by:

$$\frac{dP_n(t)}{dt} = -\left(\lambda(N-n)+\mu\right)P_n(t)+\lambda(N-n+1)P_{n-1}(t)+\mu P_{n+1}(t),$$

$$1 \leq n < N$$

$$\frac{dP_N(t)}{dt} = -\mu P_N(t)+\lambda P_{N-1}(t)$$

$$\frac{dP_0(t)}{dt} = -\lambda N P_0(t)+\mu P_1(t)$$

where $P_n(t) = P(X_t = n)$. The corresponding balance equations are given by:

$$-\mu N + \lambda_{N-1} = 0$$

$$-\left(\lambda(N-n)+\mu\right)p_n+\lambda(N-n+1)p_{n-1}+\mu_{n+1}=0,\ 1 \leq n < N$$

$$-\lambda N p_0 + \mu_1 = 0$$

where $p_n = \lim_{t \to \infty} P\,(for\,X_t = n)$. From the first equations, we can obtain

$$p_{N-1} = \left(\frac{\mu}{\lambda}\right)p_N$$

with the second equation with $n = N - 1$ we have

$$p_{N-2} = \frac{1}{2!}\left(\frac{\mu}{\lambda}\right)^2 p_N$$

In general, we get after some computation

$$p_{N-n} = \frac{1}{n!}\left(\frac{\mu}{\lambda}\right)^n p_N,\ 0 \leq n < N$$

with

$$\sum_{n=0}^{N} p_n = 1$$

then

$$p_{N-n} = \frac{\frac{1}{n!}\left(\frac{\mu}{\lambda}\right)^n}{\sum_{k=0}^{N}\frac{\left(\frac{\mu}{\lambda}\right)^k}{k!}},\ \text{if } 0 \leq n \leq N$$

and therefore

$$p_0 = \lim_{t \to \infty} P\,(X_t = 0)$$

$$= \frac{\frac{1}{N!}\left(\frac{\mu}{\lambda}\right)^N}{\sum_{k=0}^{N}\frac{\left(\frac{\mu}{\lambda}\right)^k}{k!}}$$

In particular, if $N = 5$ and $\lambda = 0.5$, $\mu = 0.3$ then $\rho_0 = 3.5564 \times 10^{-4}$. If $N = 5$ and $\lambda = 0.3$, $\mu = 0.5$ then $\rho_0 = 0.02039$.

Example 3.6 Consider an M/M/s queue model. Let

$$b_0 := 1$$

$$b_j := \frac{\lambda_0 \lambda_1 \cdots \lambda_{j-1}}{\mu_1 \mu_2 \cdots \mu_j}, \; j \geq 1$$

The process has positive recurrent if

$$\sum_{j=1}^{\infty} b_j < \infty$$

and if

$$\sum_{j=0}^{s-1} \frac{\lambda^j}{j! \mu^j} + \sum_{j=s}^{\infty} \frac{\lambda^j}{s! s^{j-s} \mu^j} < \infty$$

The first and second sum are finite, if

$$r := \frac{\lambda}{s\mu} < 1$$

in this case, we have that

$$\sum_{j=0}^{s-1} \frac{\lambda^j}{j! \mu^j} + \sum_{j=s}^{\infty} \frac{\lambda^j}{s! s^{j-s} \mu^j} = \sum_{j=0}^{s-1} \frac{\lambda^j}{j! \mu^j} + \frac{s^s r^s}{s!(1-r)}$$

and the stationary limit distribution is given by:

$$p_j = \begin{cases} p_0 \left(\frac{\lambda}{\mu}\right)^j \frac{1}{j!} & \text{if } j \leq s \\ p_0 \left(\frac{\lambda}{\mu}\right)^j \frac{1}{s! s^{j-s}} & \text{if } j \geq s \end{cases}$$

Let $(X_t)_{t \geq 0}$ a birth and death process with $\mu_i = 0$ and $\lambda_i = \lambda$ for all $i = 0, 1, 2, \ldots$ where $\lambda > 0$ is constant.

Suppose that

$$P_n(t) := p_{0n}(t) = P(X_t = n \mid X_0 = 0), \; n \geq 0$$

and let

$$P(z, t) := \sum_{n=0}^{\infty} P_n(t) z^n$$

With Kolmogorov's equations we have:

$$\frac{d P_0(t)}{dt} = -\lambda P_0(t)$$

$$\frac{d P_n(t)}{dt} = -\lambda P_n(t) + \lambda P_{n-1}(t), \ n \geq 1$$

therefore,

$$\frac{d P(z, t)}{dt} = -\lambda P(z, t) + \lambda z P(z, t)$$

$$= -\lambda (1 - z) P(z, t)$$

Since the solution to an equation of the form $\frac{dy(x)}{dx} = ay(x)$ is of the form $y(x) = C \exp(ax)$ then

$$P(z, t) = C \exp(-\lambda(1 - z)t)$$

we have

$$P(z, 0) = 1$$

with

$$P_0(0) = 1 \text{ and } P_n(0) = 0 \text{ for all } n \geq 1$$

then

$$P(z, t) = \exp(-\lambda(1 - z)t)$$

$$= \exp(-\lambda t) \sum_{n=0}^{\infty} \frac{(\lambda t)^n}{n!} z^n$$

$$= \sum_{n=0}^{\infty} \exp(-\lambda t) \frac{(\lambda t)^n}{n!} z^n$$

We have

$$P_n(t) = \exp(-\lambda t) \frac{(\lambda t)^n}{n!}, \ n \geq 0$$

We conclude that X_t has a Poisson distribution with parameter λ, and we write as

$$p_{ij}(t) = P(X_t = j \mid X_0 = i)$$

$$= \begin{cases} e^{-\lambda t} \frac{(\lambda t)^{j-i}}{(j-i)!} & \text{if } j \geq i \\ 0 & \text{if } j < i \end{cases}$$

The birth and death process with rates given by $\mu_i = 0$ and $\lambda_i = \lambda$ for all $i = 0, 1, 2, \ldots$ where $\lambda > 0$ constant is a simple Poisson process. The infinitesimal generator of this process is given by:

$$Q = \begin{pmatrix} -\lambda & \lambda & 0 & 0 & \cdots \\ 0 & -\lambda & \lambda & 0 & \cdots \\ 0 & 0 & -\lambda & \lambda & \cdots \\ 0 & 0 & 0 & -\lambda & \cdots \\ \vdots & \vdots & \vdots & \vdots & \vdots \end{pmatrix}$$

The only possible transitions in a small interval h are i a $i + 1$ or i to i. and satisfies:

$$P\left(X_{t+h} = i + 1 \mid X_t = i\right) = \lambda h + o\left(h\right)$$
$$P\left(X_{t+h} = i \mid X_t = i\right) = 1 - \lambda h + o\left(h\right)$$
$$P\left(X_{t+h} = j \mid X_t = i\right) = o\left(h\right) \text{ for all } j \geq 2$$

With

$$P\left(\text{occurs exactly one event in the interval } (0, h]\right)$$
$$= P\left(X_h = 1 \mid X_0 = 0\right)$$
$$= \lambda h + o\left(h\right)$$
$$P\left(\text{no event occurs in the interval } (0, h]\right)$$
$$= P\left(X_h = 0 \mid X_0 = 0\right)$$
$$= 1 - \lambda h + o\left(h\right)$$
$$P\left(\text{more than one event occurs in the interval } (0, h]\right)$$
$$= P\left(X_h = j \mid X_0 = 0\right)$$
$$= o\left(h\right) \text{ for all } j \geq 2$$

We know that $\{X_t; t \geq 0\}$ is a Poisson process with parameter λt, as seen in the previous chapter.

Yule process

Let $\{X_t; t \geq 0\}$ a birth and death process with $\lambda_i = i\lambda$ and $\mu_i = 0$ for all $i = 0, 1, \ldots$

We have the following Kolmogorov equations

$$\frac{d p_{ij}\left(t\right)}{dt} = -j\lambda p_{ij}\left(t\right) + (j - 1)\lambda p_{i,j-1}\left(t\right)$$

where

$$p_{ij}\left(t\right) = P\left(X_t = j \mid X_0 = i\right)$$

Define the probability generating function

$$P_i(z, t) := \sum_{j=0}^{\infty} P_{ij}(t) z^j$$

Differentiating with respect to t and having that $p_{ij}(t) = 0$ if $j < i$ we have

$$\frac{dP_i(z, t)}{dt} = -\lambda \sum_{j=i}^{\infty} j p_{ij}(t) z^j + \lambda \sum_{j=i+1}^{\infty} (j-1) p_{i,j-1}(t) z^j$$

$$= -\lambda z \sum_{j=i}^{\infty} j p_{ij}(t) z^{j-1} + \lambda \sum_{l=i}^{\infty} l p_{il}(t) z^{l+1}$$

$$= -\lambda z \sum_{j=i}^{\infty} j p_{ij}(t) z^{j-1} + \lambda z^2 \sum_{l=i}^{\infty} l p_{il}(t) z^{l-1}$$

$$= -\lambda z \frac{dP_i(z, t)}{dt} + \lambda z^2 \frac{dP_i(z, t)}{dt}$$

We proceed to obtain the solution of the above equation using the variable separable method with variables t and z. Suppose that the solution denoted by $P(z, t)$ (we omit the i for convenience) is of the form

$$P(z, t) = h(z) g(t)$$

With initial conditions

$$P(z, 0) = \sum_{j=0}^{\infty} P_{ij}(0) z^j = z^i$$

$$P(1, t) = \sum_{j=0}^{\infty} P_{ij}(t) = 1$$

Differentiating with respect to t we have

$$\frac{dP(z, t)}{dt} = h(z) \frac{dg(t)}{dt}$$

where

$$h(z) \frac{dg(t)}{dt} = -\lambda z \frac{dh(z)}{dz} g(t) + \lambda z^2 \frac{dh(z)}{dz} g(t)$$

dividing by $h(z) g(t)$ we get

$$\frac{\frac{dg(t)}{dt}}{g(t)} = -\lambda z \frac{\frac{dh(z)}{dz}}{h(z)} + \lambda z^2 \frac{\frac{dh(z)}{dz}}{h(z)}$$

for these expressions to be equal, they must be equal one by one constant, we say the constant is α, then we have:

$$\frac{dg\,(t)}{dt} = \alpha g\,(t)$$

solving

$$g\,(t) = \beta \exp(\alpha t)$$

with β constant. therefore,

$$\frac{\frac{dh(z)}{dz}}{h\,(z)} = \frac{\alpha}{\lambda z^2 - \lambda z}$$

$$= \frac{\alpha}{\lambda}\left[\frac{1}{z-1} - \frac{1}{z}\right],\ z \neq 0, 1$$

then,

$$\ln h\,(z) = \frac{\alpha}{\lambda}\left[\ln|z-1| - \ln|z|\right]$$

$$h\,(z) = \left(\frac{1-z}{z}\right)^{\frac{\alpha}{\lambda}},\ z \neq 0$$

Since the general solution of the equation is of the form:

$$P\,(z, t) = G\,(h\,(z)\,g\,(t))$$

$$= G\left(\beta \exp(\alpha t)\left(\frac{1-z}{z}\right)^{\frac{\alpha}{\lambda}}\right)$$

and since:

$$G\left(\beta\left(\frac{1-z}{z}\right)^{\frac{\alpha}{\lambda}}\right) = P\,(z, 0) = z^i$$

by taking

$$u = \beta\left(\frac{1-z}{z}\right)^{\frac{\alpha}{\lambda}}$$

we get

$$z = \left[\left(\frac{u}{\beta}\right)^{\frac{\lambda}{\alpha}} + 1\right]^{-1}$$

and

$$G\,(u) = \left[\left(\frac{u}{\beta}\right)^{\frac{\lambda}{\alpha}} + 1\right]^{-i}$$

therefore we can obtain

$$P(z, t) = G\left(\beta \exp(\alpha t)\left(\frac{1-z}{z}\right)^{\frac{\alpha}{\lambda}}\right)$$

$$= \left(\left[\exp(\alpha t)\left(\frac{1-z}{z}\right)^{\frac{\alpha}{\lambda}}\right]^{\frac{\lambda}{\alpha}} + 1\right)^{-i}$$

$$= \frac{1}{\left[\left(\frac{1-z}{z}\right)\exp(\lambda t) + 1\right]^{i}}$$

$$= \left[\frac{z\exp(-\lambda t)}{1 - z(1 - \exp(-\lambda t))}\right]^{i}$$

the above expression corresponds to the probability generating function of a random variable with negative binomial distribution with parameters i and $\exp(-\lambda t)$.

3.3   COVID-19 Modeling

This section presents an example of COVID-19 infection modeling using a continuous-time Markov chain modeling. One of the main concerns in epidemiological models is to obtain the transition probabilities and the expected number of people infected during the COVID-19 pandemic.

Let $X(t)$ be the number of people infected in an arbitrary time t. Suppose that initially, there are n_0 people infected with the COVID-19 virus, for some $n_0 > 0$, the Kolmogorov differential equations can be written as the following:

$$P'_{i0}(t) = \lambda_0 P_{i,0}(t) + \mu_j P_{i,1}(t)$$

$$P'_{ij}(t) = \lambda_{j-1} P_{i,j-1}(t) - (\lambda_j + \mu_j) P_{ij}(t) + \mu_{j+1} P_{i,j+1}(t)$$

with some initial condition $P_{ij}(0) = \delta_{i,j}$.

In this model, we assume that t λ_j and μ_j are constant, and for the simplicity $\lambda_j = \lambda$ and $\mu_j = \mu$ for all j.

We follow the approach presented in [3] for calculating transition probabilities of a birth and death Markov process based on the matrix method to obtain mathematical expectation of the number of infected individuals after time t. The movement of the state j must be to the state $j-1$ or to the state $j+1$. To simplify, we suppose the transition to the state j is independent of the initial state. Therefore, the equations above can be written as:

$$P_1'(t) = -\lambda P_1(t) + \mu P_2(t)$$

$$P_j'(t) = \lambda P_{j-1}(t) - (\lambda + \mu) P_j(t) + \mu P_{j+1}(t) \qquad 1 \le j \le n$$

Consequently, we obtain the following:

$$P_1'(t) = -\lambda P_1(t) + \mu P_2(t)$$

$$P_2'(t) = \lambda P_1(t) - (\lambda + \mu) P_2(t) + \mu P_3(t)$$

$$\cdot$$
$$\cdot$$
$$\cdot$$

$$P_n'(t) = \lambda P_{n-1}(t) - \mu P_n(t)$$

Then the matrix form the equations can be written as $P' = AP$, where A has the following form:

$$A := \begin{bmatrix} -\lambda & \mu & 0 & 0 \cdots 0 \\ \lambda & -(\lambda + \mu) & \mu & 0 \cdots 0 \\ 0 & \lambda & -(\lambda + \mu) & 0 \cdots 0 \\ \vdots & & & \\ 0 & 0 & \cdots & 0 \ \lambda \ \mu \end{bmatrix}$$

and the vectors P and P' are:

$$P = \begin{bmatrix} P_1(t) \\ P_2(t) \\ \vdots P_n(t) \end{bmatrix} \quad and \ P' = \begin{bmatrix} P_1'(t) \\ P_2'(t) \\ \vdots P_n'(t) \end{bmatrix}$$

The solution of the system of the equation has the form:

$$P(t) = e^{At} P(0) \tag{3.12}$$

where $P(0)$ are the initial conditions. Let $M(t)$ be the average number of infected individuals at time t, we have

$$M(t) = E[X(t)] = \sum_{j=1} j P_j(t) \tag{3.13}$$

After some computation of solving the system of differential equations, we get

$$M(t) = (\lambda - \mu)t + n_0 \tag{3.14}$$

Let us consider 1000 individuals with initial infected 50 individuals at the time $t = 0$. The probability of transmission of the infection with rate $\lambda = 0.04$. Assume that the rate recovery $\mu_1 = 0.07$, and the death rate $\mu_2 = 0.01$, and $\mu = \mu_1 + \mu_2 = 0.08$ respectively (see [3]) Taking into account the information, a simulation with three trajectories shows the behavior of the individuals in a time interval between 0 and 80. Finally, we resume in Table 3.1, the expected value M(t) of the number of infected is calculated for a time interval between 0 and 80 using the Eq. 3.13 and is compared with the number of infected N(t) of the simulated trajectory.

```python
import numpy as np
import matplotlib.pyplot as plt
import itertools as it
%matplotlib inline

def simplebirthdeath(N, samplepaths, b, d):
    X0 = 36 # initial population size
    s = np.zeros(N)
    for k in range(N):
        s[k] = k
    X = np.zeros((samplepaths, N))
    X[:,0] = X0
    for j in range(samplepaths):
        i = 0
        while X[j,i] > 0 and i < (N-1):
            U1 = np.random.rand()
            U2 = np.random.rand()
            h = - np.log(U1)/((b+d)*X[j,i])
            if U2 < b/(b+d):
                X[j,i+1] = X[j,i] + 1 # birth occurs
            else:
                X[j,i+1] = X[j,i] - 1 # death occurs
            i += 1
    return [s,X]


maxPop = 1000
walkers = 5
birth = 0.04
death = 0.08

X = simplebirthdeath(maxPop, walkers, birth, death)
```

Table 3.1 M(t) is the number of infected individuals and N(t) is the trajectory which is simulated

t	0	5	10	15	20	25	30	35	40	45	50
N(t)	36.	35.	32.	29.	30.	31.	30.	31.	30.	27.	22.
M(t)	36.	35	34	33	32	31	30	29	28	27	26

```
X0 = X[0]
fig, ax = plt.subplots()
for i in range(3):
  plt.step(X0, X[1][i])
  plt.axis([0, 80, 0, 40])
  ax.set_xlabel("Time", fontsize=14)
  ax.set_ylabel("Population Size", fontsize=14)
  plt.xticks(fontsize=14)
  plt.yticks(fontsize=14)
  plt.tight_layout()
  plt.savefig("1")

X0 = X[0]
fig, ax = plt.subplots()
for i in range(3):
  plt.step(X0, list(it.accumulate(X[1][i])))
  plt.axis([0, 80, 0, 1800])
  ax.set_xlabel("Time", fontsize=14)
  ax.set_ylabel("Total Population Size", fontsize=14)
  plt.xticks(fontsize=14)
  plt.yticks(fontsize=14)
  plt.tight_layout()
  plt.savefig("2")
```

We have given an illustrative example of COVID-19 through a simulation with five trajectories to show the behavior of the number of infected individuals in the time interval (Figs. 3.1 and 3.2). This modeling idea gives an overview of a birth and death process applied to epidemiology. The expected value obtained is very close to the number generated by the processes. The proposed approach opens the way to explore the applications further and extends to include factors such as prevention and control measures for public health.

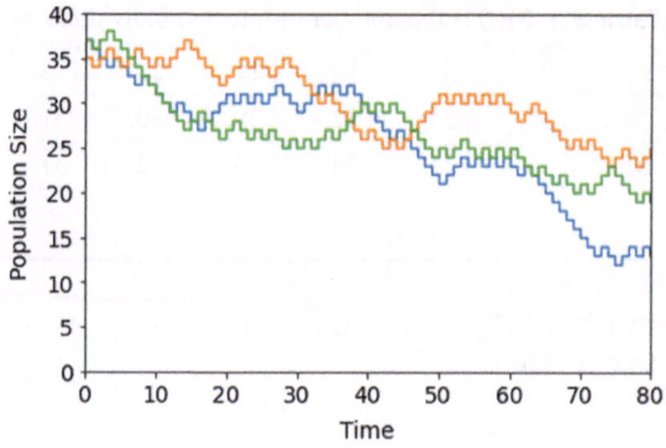

Fig. 3.1 The number of infected individuals in the time interval $0 \leq t \leq 80$

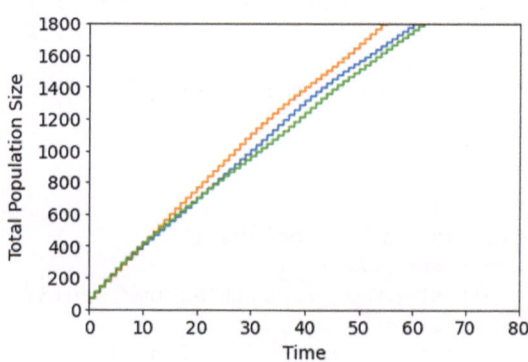

Fig. 3.2 The cumulative number of infected individuals in the time interval $0 \leq t \leq 80$

References

1. Todorovic, P. (1992). *An introduction to stochastic processes and their applications*. New York: Springer.
2. Castañeda, L. B., Arunachalam, V., & Dharmaraja, S. (2012). *Introduction to probability and stochastic processes with applications* (1st ed.). New Jersey: Wiley.
3. Eslahchi, C., & Movahedi, F. (2012). Calculation of transition probabilities in the birth and death Markov process in the epidemic model. *Mathematical and Computer Modelling, 55*(3–4), 810–815.

Branching Processes

4

The theory of branching processes began during the second half of the 19th century from the approach of a problem related to the extinction of aristocratic surnames. The theory offers a mathematical model for studying populations in which individuals live, reproduce, and die independently of one another. The first indications of proposed problems related to branching processes are found in the 18th century in the famous three-volume book by researcher Thomas Malthus, entitled *Essay on the Principles of Populations*, which state that an uncontrolled population must grow exponentially. Malthus also related that, in a town in Berne, of the 487 existing bourgeois families, 379 became extinct in the span of two centuries (1583–1783); however, he did not study the reason for this apparent paradox. The first to try to explain the problem was the French mathematician Bienayme (1796–1878). Unfortunately, there is no written record that he could correctly relate the probability of extinction to the average number of male children of each individual.

In the Galton-Watson process, the lifespan of each individual is a unit of time. A natural generalization is to allow such lifetimes to be random variables. These processes are not, in general, Markovian unless the life lengths of the individuals are independent and equally distributed random variables with an exponential distribution. In the model considered, independence from the past of the process is assumed. However, in various situations, such as the case of the description of the spread of a contagious disease, the history of the process is very important because it provides information about the control measures adopted. In such cases, the behavior of the process is described using a one-type branching process dependent on the size of the population.

4.1   Galton-Watson Process

The simplest branching process is called a Galton-Watson process. This process has been used to describe the spread of an epidemic: suppose that in the initial time, an individual is carrying a virus that can infect other individuals only during a certain period of time, after which it becomes inactive. Individuals contagious by this initial individual begin to spread the virus independently of each other, in the same way as their parent, and so on. Consequently, the number Z_n of infected individuals at time $n \geq 1$ is equal to:

$$Z_n = \sum_{k=1}^{Z_{n-1}} X_{n,k}$$

where $X_{n,k}$ represents the number of individuals infected by the kth individual present at time $(n-1)$.

Let $m := E(Z_1)$, the number of new infections in the individual population. There are three possibilities $m < 1$, $m = 1$ or $m > 1$, respectively, subcritical, critical and supercritical cases. In the case where $m < 1$ we have that with probability 1, Z_n tends to 0 when $n \to \infty$. If $m > 1$, the process is explosive, and the number of infected individuals grows exponentially. If $m = 1$, the number of infected individuals does not change, on average, and keeps fluctuating between the subcritical and critical cases. Therefore, when an epidemic arises, such as COVID-19, measures must be taken so that m is less than 1, as used to measure the concept of flattening the contagion curve.

Definition 4.1 A Galton-Watson process $(Z_n)_{n \in \mathbb{N}}$ with offspring distribution $(p_k)_{k \geq 0}$ is a discrete-time homogeneous Markov chain taking values in the set $\mathbb{Z}_+$ of non-negative integers. Its transition probabilities p_{ij} with $i, j \in \mathbb{N}$ are expressed in terms $\{p_k, k = 0, 1, 2, ..\}$, $p_k \geq 0$ and $\sum_k p_k = 1$ by:

$$p_{ij} := P(Z_{n+1} = j \mid Z_n = i) = \begin{cases} p_j^{*i}, & i \geq 1, j \geq 0 \\ \delta_{0j}, & i = 0, j \geq 0 \end{cases}$$

where δ_{ij} is the Kronecker delta and p_j^{*i}, $j = 0, 1, 2, \ldots$ is the jth component of the ith convolution of $\{p_k, k = 0, 1, 2, \ldots\}$.

To avoid trivialities, it is assumed that $p_0 + p_1 < 1$ and that $p_j \neq 1$ for all j, otherwise the process extinct with probability 1. Unless the contrary is stated, we assume that $Z_0 = 1$. It further follows from the definition of the Galton-Watson process that the distribution of $(Z_{n+1}, Z_{n+2}, \ldots, Z_{n+k}, \ldots)$ given that $Z_n = i$ is equal to the distribution of a sum of i independent copies of $(Z_1, Z_2, \ldots, Z_k, \ldots)$.

Example 4.1 Galton-Watson process with binary reproduction Let $(Z_n)_{n\in\mathbb{N}}$ a Galton-Watson process with offspring distribution $(p_k)_{k\geq0}$ given by:

$$0 < p_0 < 1, \quad p_0 + p_2 = 1,$$
$$p_k = 0, \text{ for all } k \neq 0, 2$$

The offspring tree for the binary reproduction in Fig. 4.1.

Example 4.2 Reproduction of DNA segments of length N

Let $(Z_n)_{n\in\mathbb{N}}$ a Galton-Watson process with offspring distribution $(p_k)_{k\geq0}$ given by

$$p_0 := \alpha$$
$$p_1 := (1 - \alpha)\left(1 - \beta^N\right)$$
$$p_2 := (1 - \alpha)\beta^N$$
$$p_k := 0, \text{ for all } k \geq 3$$

where α and β are constants with $\alpha, \beta \in (0, 1)$.

Example 4.3 Binomial reproduction

Let $(Z_n)_{n\in\mathbb{N}}$ a Galton-Watson process with offspring distribution $(p_k)_{k\geq0}$ given by:

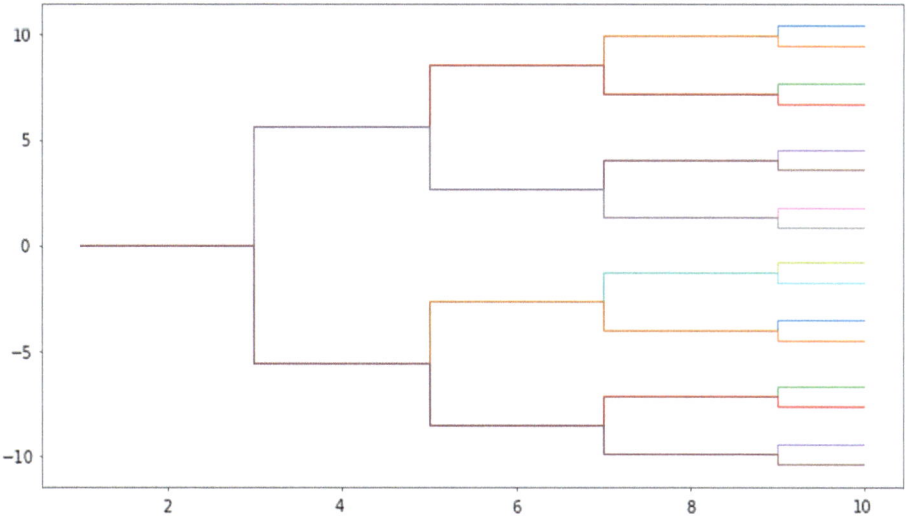

Fig. 4.1 Offspring tree for the binary reproduction

$$p_k := \binom{m}{k} \alpha^k (1 - \alpha)^{m-k} ; \ k = 0, 1, 2, \ldots, m$$

$$p_k := 0, \ k \notin \{0, 1, 2, \ldots, m\}$$

where m is a natural number greater than 0 and $\alpha \in (0, 1)$.

Example 4.4 Geometric reproduction Let $(Z_n)_{n \in \mathbb{N}}$ a Galton-Watson process with offspring distribution $(p_k)_{k \geq 0}$ given by:

$$p_k := \alpha \, (1 - \alpha)^k , \ k \geq 0$$

where $\alpha \in (0, 1)$.

Example 4.5 Lotka process Let $(Z_n)_{n \in \mathbb{N}}$ a Galton-Watson process offspring distribution $(p_k)_{k \geq 0}$ given by:

$$p_0 := \beta$$

$$p_k := (1 - \beta) \alpha \, (1 - \alpha)^{k-1} , \ k \geq 1$$

where $\alpha, \beta \in (0, 1)$.

In the following results and some important theorems, we recommend the reader refer to Athreya and Ney [1]. For convenience, we define the notation of the probability-generating function of the random variable Z_1 is denoted by $f(s)$ and is given by

$$f(s) := E\left(s^{Z_1}\right)$$

Example 4.6 The binary reproduction is given by:

$$f(s) = p_0 + p_2 s^2$$

Example 4.7 If $(Z_n)_{n \in \mathbb{N}}$ is a Galton-Watson process with offspring distribution given by:

$$p_k := \begin{cases} \frac{1}{3} & \text{if } k = 0, 1, 2 \\ 0 & \text{otherwise} \end{cases}$$

then

$$f(s) = \frac{1}{3} \left(1 + s + s^2\right)$$

Example 4.8 In the binomial reproduction model with parameters m and $0 < \alpha < 1$ we have:

$$f(s) = (\alpha s + (1 - \alpha))^m$$

Example 4.9 In the model with offspring distribution of geometric with parameter α we obtain

$$f(s) = \frac{\alpha}{1 - (1 - \alpha)s}$$

Example 4.10 In the Lotka process, the probability-generating function of Z_1 is:

$$f(s) = \beta + (1 - \beta) \frac{\alpha s}{1 - (1 - \alpha)s}$$

Theorem 4.1 Let $f(s) := E[s^{Z_1}]$ the probability-generating function of Z_1. Then we have that

a. $f_n(s) := E[s^{Z_n}] = f_{n-1}(f(s))$, that is the probability generating function of Z_n is the nth iteration of f.
b. $E(u^{Z_n} v^{Z_{n+k}}) = f_n(u f_k(v))$, that is, the joint probability generating function of Z_n and Z_{n+k} is given by $f_n(u f_k(v))$.

Example 4.11 In the Galton-Watson process with geometric offspring distribution, we have:

$$f_n(s) = \begin{cases} \frac{m^n(1-s)+ms-1}{m^{n+1}(1-s)+ms-1} & \text{if } \alpha \neq \frac{1}{2} \\ \frac{n(1-s)+s}{n+1-ns} & \text{if } \alpha = \frac{1}{2} \end{cases}$$

where $m = \frac{1-\alpha}{\alpha}$.

Theorem 4.2 Let $m := E(Z_1)$ and $\sigma^2 := Var(Z_1)$. Then:

a. $E(Z_n) = m^n$
b.
$$Var(Z_n) = \begin{cases} \sigma^2 \frac{m^n(m^n-1)}{m(m-1)}, & \text{if } m \neq 1 \\ \sigma^2 n, & \text{if } m = 1 \end{cases}$$

c. $Cov(Z_n, Z_{n+k}) = m^k Var(Z_n)$.

Example 4.12 If $(Z_n)_{n \in \mathbb{N}}$ is a Galton-Watson process with offspring distribution given by:

$$p_k := \begin{cases} \frac{1}{3} & \text{if } k = 0, 1, 2 \\ 0 & \text{otherwise} \end{cases}$$

then

$$m = E(Z_1) = 1$$

and

$$E\left(Z_n\right) = 1$$

$$Var\left(Z_n\right) = \frac{2}{3}n$$

In the previous theorem, we observed that:

1. If $m < 1$ then $\lim_{n\to\infty} E\left(Z_n\right) = 0$ and $\lim_{n\to\infty} Var\left(Z_n\right) = 0$
2. If $m = 1$ then $E\left(Z_n\right) = 1$
3. If $m > 1$ then $\lim_{n\to\infty} E\left(Z_n\right) = \infty$ and $\lim_{n\to\infty} Var\left(Z_n\right) = \infty$

Therefore, if the Galton-Watson process describes the spread of an epidemic, then the ideal scenario is that $m < 1$ because, in the long run, with probability 1, the number of infected is zero. In the case of an epidemic, it is advisable to take measures in order to achieve that $m < 1$.

Remark 4.1 Due to the fact that $(Z_{n+1}, Z_{n+2}, \ldots, Z_{n+k}, \ldots)$ given that $Z_n = i$ is equal to the distribution of a sum of i independent copies of $(Z_1, Z_2, \ldots, Z_k, \ldots)$ then

$$E\left(s^{Z_n} \mid Z_0 = i\right) = \left(E\left(s^{Z_n} \mid Z_0 = 1\right)\right)^i = \left(f_n\left(s\right)\right)^i, \text{ for } i \geq 1$$

consequently,

$$P\left(Z_n = j \mid Z_0 = i\right) = \frac{1}{j!} \frac{\partial^j}{\partial s^j} \left(f_n\left(s\right)\right)^i \mid_{s=0}$$

Remark 4.2 Let $s \in \mathbb{R}$. From the definition of f as a power series with non-negative coefficients $p_k, k = 0, 1, 2, ..$ whose sum is 1 and such that $p_0 + p_1 < 1$, it follows that:

a. f is strictly increasing and convex at $[0, 1]$.
b. $f(0) = p_0$ and $f(1) = 1$
c. If $m \leq 1$ then $f(s) > s$ for all $s \in [0, 1)$
d. If $m > 1$ then $f(s) = s$ has a unique root in $[0, 1)$.

Theorem 4.3

a. $\lim_{n\to\infty} f_n(s) = q$, for all $s \in [0, 1)$, being q the smallest solution of the equation $f(s) = s$.
b. $q = P(Z_n \to 0)$, that is, q is the probability of extinction of the process.
c. $m \leq 1$, if and only if, $q = 1$.

Example 4.13 Binary reproduction The probability of extinction of the process is the smallest root q of the equation

$$p_2 s^2 - s + p_0 = 0$$

that is,

$$q = \frac{p_0}{p_2}$$

If, for example, $p_0 = p_2 = \frac{1}{2}$ then $q = 1$.

Example 4.14 In the binomial reproduction model with parameters $m = 5$ and $\alpha = \frac{1}{4}$ we have that q is the smallest root of the equation

$$\left(\frac{1}{4}s + \frac{3}{4}\right)^5 = s$$

$$s^5 + 15s^4 + 90s^3 + 270s^2 - 619s + 243 = 0$$

That is $q \simeq 0.553$.

Theorem 4.4 *Let* $(Z_n)_{n \in \mathbb{N}}$ *be a Galton-Watson process whose offspring distribution* $(p_k)_{k \geq 0}$ *has mean* μ *and variance* σ^2. *Then*

1. *For each* $k \geq 0$ *it is satisfied that:*

$$kP\,(Z_n > 0 \mid Z_0 = 1)\,[P\,(Z_n = 0 \mid Z_0 = 1)]^{k-1} \leq P\,(Z_n > 0 \mid Z_0 = k) \leq kP\,(Z_n > 0 \mid Z_0 = 1)$$

2. *If* $m < 1$ *then, when* $n \to \infty$ *we have:*

$$k\,(1 - m)\,\frac{m^{n+1}}{\sigma^2} \leq P\,(Z_n > 0 \mid Z_0 = k) = P\,(T_e > n \mid Z_0 = k) = km^n$$

   where $T_e := \inf\{n \in \mathbb{N} : Z_n = 0\}$, *this is* T_e *represents the time of extinction of the process.*

Proof See Roelly [2].   □

The following theorem states that if a population follows according to a Galton-Watson process, then it must either go extinct or explode with probability 1.

Theorem 4.5 *Let* $(Z_n)_{n \in \mathbb{N}}$ *be a Galton-Watson process. Then:*

$$P(Z_n \to 0) + P(Z_n \to \infty) = 1$$

Proof All states are transient for $k \geq 1$, that is

$$P(Z_{n+i} \neq k, i \geq 1 \mid Z_n = k) > 0$$

Indeed

$$P(Z_{n+i} \neq k, i \geq 1 \mid Z_n = k) \geq P(Z_{n+1} \neq k \mid Z_n =, k)$$

$$\geq \begin{cases} 1 - P(Z_{n+1} = k \mid Z_n = k), \text{ if } p_0 = 0 \\ P(Z_{n+1} = 0 \mid Z_n = k), \text{ if } p_0 > 0 \end{cases}$$

$$= \begin{cases} (1 - p_1)^k, \text{ if } p_0 = 0 \\ p_0^k, \text{ if } p_0 > 0 \end{cases}$$

$$> 0$$

Therefore, with probability 1, we have $Z_n \to 0$ or $Z_n \to \infty$, and that the stationary distribution π is the trivial distribution, that is,

$$\pi_0 = 1$$
$$\pi_k = 0 \text{ for all } k \geq 1$$

Definition 4.2 Let $(Z_n)_{n \in \mathbb{N}}$ be a Galton-Watson process.

a. If $m > 1$ then the process is called *supercritical*.
b. If $m = 1$ then the process is called *critical*.
c. If $m < 1$ then the process is called *subcritical*.

For a supercritical Galton-Watson process it is satisfied that $P(Z_n \to \infty) > 0$

Example 4.15 According to the International Union for the Conservation of Nature (IUCN), the survival of the North Atlantic black whale (or North Atlantic right whale) is critically endangered and could disappear if no action is taken urgently to preserve it. In the year 1992 there were 295 living whales, and from 1990 to 2010 the number of living whales increased on average by 2.8% per year. Thus, it is estimated that in 1994 there were approximately 312 copies. In its report dated October 24, 2022 [3], the Whale Consortium North Atlantic Francas (NARWC) announced that the population of North Atlantic right whales, critically endangered extinction, now at 340, down from 348, continues to decline for a decade. The species' population has plummeted by a 30% in the last decade, down from 481 in 2011.

The law of reproduction of the black whales of the North Atlantic is given by:

$$r_0 = b, r_1 = (1 - b)(1 - a); r_2 = (1 - b)a$$

where b is the probability that a female whale will die in the next year and a is the probability that a whale will give birth to a small female whale in the next year. In the year 1994, the probability b was estimated at 0.06 and the probability a was estimated at 0.038, and it was also determined that the population of female whales was approximately 150 specimens [4].

Let $(Z_n)_{n\geq 1}$ be the process with

$$Z_n := \text{"number of female North Atlantic black whales in the } n\text{th year"}$$

The process $(Z_n)_{n\geq 1}$ is a Galton-Watson process. With use of the data mentioned, we obtain

$$m := E(Z_1) = r_1 + 2r_2 \approx 0.976$$

and

$$\sigma^2 := Var(Z_1) = r_1 + 4r_2 - m^2 \approx 0.095$$

since $m < 1$ then

$$k(1-m)\frac{m^{n+1}}{\sigma^2} \leq P(Z_n > 0 \mid Z_0 = k) = P(T_e > n \mid Z_0 = k) \leq km^n$$

where T_e represents the time of extinction of the population. If environmental conditions do not change, then the population of whales will become extinct within n years with a probability greater than 0.99, if $km^n \leq 0.01$, that is,

$$150 \times (0.976)^n \leq 0.01 \iff n \geq 395$$

This means that in the year 2389 with a probability greater than 0.99 there will be no more female black whales in the North Atlantic. Unfortunately, the living conditions of the whales have not been maintained in the last decades. The increase in the temperature of the oceans forces the whales to move in search of better environmental conditions, which increases the chances of colliding with fishing boats and being stranded and entangled in the devices deployed by the fishermen. In [4] the scientists expressed concern about the downward trend in the number of whale females capable of reproducing. Research has also found troubling evidence of the decline in the size of the body, partly due to frequent entanglements in fishing gear, and smaller female right whales produce fewer calves, On the other hand, when whales become entangled in fishing gear, may drown immediately or die later due to injuries caused, tanglesalso influence the reproductive capacity of whales, it is currently estimated that the time between the calves of the whale is almost ten years old. For the reasons stated, the extinction of the North Atlantic whales may be very close before the estimate given above.

4.2 Multi-type Galton-Watson Process

A possible generalization of the simple Galton-Watson process is to consider processes involving several different types of individuals. For example, in the reproduction of a specific type of bacterium, the original bacterium may give origin to a mutant form that behaves differently, or, the distribution of offspring may depend on the age group to which the individual belongs.

Let us assume that there are $k \in \mathbb{N}$ different types(k fixed). Since the considered population consists of individuals that can belong to k different types, then the numerical development of the population must be described by a vector $\vec{Z}_n := (Z_n^{(1)}, Z_n^{(2)}, \ldots, Z_n^{(k)})$, where $Z_n^{(j)} :=$ "number of individuals of the nth generation belonging to the type j", where $j = 1, 2, \ldots, k$ and $n \in \mathbb{N}$. As in the case of the simple Galton-Watson process, it is assumed that individuals reproduce independently of each other and independently of the history of the process. The distribution of offspring obviously depends on the type the individual.

Definition 4.3 The multitype Galton-Watson process $(\vec{Z}_n)_{n \in \mathbb{N}}$ is a Markov chain with state space $\mathbb{N}^k$ such that:

a. $\vec{Z}_0 = \vec{z} \in \mathbb{N}^k - \left\{ \vec{0} \right\}$ fixed.

b. The jth component of the vector $\vec{Z}_{n+1}$ is given by:

$$Z_{n+1}^j = \sum_{i=1}^{k} \sum_{l=1}^{Z_n^i} Y_l^{(n)}(i, j)$$

where the $Y_l^{(n)}(i, j)$ are independent random variables and the random vectors

$$\vec{Y}_{i,l}^{(n)} := (Y_l^{(n)}(i, 1), Y_l^{(n)}(i, 2), \ldots, Y_l^{(n)}(i, k))$$

are distributed according to the offspring distribution $p^{(i)}$ corresponding to individuals of type i. That is, $p^{(i)} := (p^{(i)}(\vec{r}))_{\vec{r} \in \mathbb{N}^k}$ where $p^{(i)}(\vec{r}) = p^{(i)}(r_1, r_2, \ldots, r_k) :=$ Probability that an individual of type i has r_1 children of type 1, r_2 children of type $2, \ldots, r_k$ children of type k.

c.

$$P(\vec{Z}_0 = \vec{t}) = \begin{cases} 1 \ if \ \vec{t} = \vec{z} \\ 0 \ otherwise \end{cases}$$

d. The transition probabilities are given by:

$$P(\vec{Z}_{n+1} = \vec{t} \mid \vec{Z}_n = \vec{r}) = \left(\underbrace{p^{(1)} * \cdots * p^{(1)}}_{r_1-times} \right) * \left(\underbrace{p^{(2)} * \cdots * p^{(2)}}_{r_2-times} \right) * \cdots *$$

$$\left(\underbrace{p^{(k)} * \cdots * p^{(k)}}_{r_k-times} \right) (t_1, t_2, \ldots, t_k)$$

$$= \left(p^{(1)*r_1} * p^{(2)*r_2} * \cdots * p^{(k)*r_k} \right) (t_1, t_2, \ldots, t_k)$$

where $\vec{r} = (r_1, r_2, \ldots, r_k)$, $\vec{t} = (t_1, t_2, \ldots, t_k) \in \mathbb{N}^k$.

Definition 4.4 If $\vec{z}_0 = e_i = (0, \ldots, 0, \underset{place-i}{1}, 0, \ldots, 0)$ then we denote the probability-generating function of $\vec{Z}_n$ by $f_n^{(i)}(\vec{s})$ with $\vec{s} \in [0, 1]^k$ and define

$$\vec{f}_n(\vec{s}) := (f_n^{(1)}(\vec{s}), f_n^{(2)}(\vec{s}), \ldots, f_n^{(k)}(\vec{s})).$$

Definition 4.5 The expected value matrix $M := (m_{ij})_{1 \le i, j \le k}$ is a defined as follows:

$$m_{ij} := E(Z_1^j \mid \vec{Z}_0 = e_i), \ 1 \le i, j \le k.$$

Definition 4.6

a. The Galton-Watson multitype process $(\vec{Z}_n)_{n \in \mathbb{N}}$ is said to be *regular positive* if $M = (m_{ij})_{1 \le i, j \le k}$ is regular positive, that is if there exists l positive integer such that $M^l > 0$

b. $(\vec{Z}_n)_{n \in \mathbb{N}}$ is said to be *singular,* if $P(Z_1^1 + \cdots + Z_1^k = 1 \mid \vec{Z}_0 = e_i) = 1$ for all $i = 1, 2, \ldots, k$; that is, when each individual has exactly one child.

Definition 4.7 Let $(\vec{Z}_n)_{n \in \mathbb{N}}$ be a multitype Galton-Watson process, we say that the process is supercritical, critical or subcritical if $\rho > 1$, $\rho = 1$ or $\rho < 1$ respectively, where ρ is the maximum simple eigenvalue of the M matrix of expected values.

In a similar way to what happens in the case of the simple Galton- Watson has that if a population develops according to a multitype process of Galton-Watson, then with probability one, it must become extinct or explode, as established by the following theorem.

Theorem 4.6 If $(\vec{Z}_n)_{n \in \mathbb{N}}$ is a Galton-Watson multi-type process, positive regular and non-singular then:

$$P(\vec{Z}_n \underset{n \to \infty}{\longrightarrow} \infty \mid \vec{Z}_0 = \vec{t}) + P(\vec{Z}_n \underset{n \to \infty}{\longrightarrow} 0 \mid \vec{Z}_0 = \vec{t}) = 1$$

for each $\vec{t} \in \mathbb{N}^k$.

Next, we will give the probability of extinction of the process $(\vec{Z}_n)_{n \in \mathbb{N}}$.

Definition 4.8

a.

$$q^{(i)} := P(\vec{Z}_n = \vec{0} \text{ for some } n \mid \vec{Z}_0 = e_i)$$

That is, $q^{(i)}$ is the probability of eventual extinction of the given process that it has been with a single particle of type i.

b.

$$\vec{q} := (q^{(1)}, q^{(2)}, \dots, q^{(k)}).$$

4.3   Continuous-Time Branching Process

In 1952, Bellman and Harris proposed generalizing Galton Watson's process. In this process, considered a discrete distribution of offspring $(p_k)_{k \geq 0}$ and instead of assuming that the life span of individuals is a unit of time, such life lengths are allowed to be a random variable. These processes are not, in general, Markovian unless the life lengths of individuals are random variables independent and equally distributed with exponential distribution. These processes are called age-dependent branching processes. In this process, the probability that an individual of age τ dies in the time interval $(\tau, \tau + d\tau)$ is, generally, a non-constant function of the age τ. A population is considered as follows: an individual born at time $t = 0$ has a lifespan l, where l is a random variable with distribution function G. At the end of that time interval, the individual is replaced by a random number of individuals of the same type of age 0, according to a distribution of offspring $(p_k)_{k \geq 0}$. Each of the children lives a random time l with $l \overset{d}{=} G$ and on death, it is replaced by a random number of children according to the distribution of offspring $(p_k)_{k \geq 0}$. Assume that the lengths of lives of individuals are independent random variables, the probabilities p_k are independent of time, of the age of the individuals at the time of being replaced, the current state, and the process history. The generation to which an individual belongs corresponds to the number of ancestors he possesses. The proof of the following results can be consulted in [5].

Theorem 4.7 *Let $(Z_t)_{t \geq 0}$ be an age-dependent process with distribution of the life length of individuals G and with distribution of offspring (p_k) $k \geq 0$ then the probability generating function $F(s, t) := E(s^{Z_t})$ satisfies the following integral equation:*

$$F(s, t) = s(1 - G(t)) + \int_0^t f(F(s, t - u)) \, dG(u)$$

where

$$f(s) := \sum_{k=0}^{\infty} p_k s^k; \ |s| \le 1$$

Remark 4.3 If $G\left(0^+\right) = 0$ and $m := f'(1) < \infty$ then $P\left(Z_t < \infty\right) = 1$ for each $t \ge 0$. The condition $G\left(0^+\right) = 0$ indicates that the probability of instant death is zero. We assume that $G\left(0^+\right) = 0$ and $m < \infty$.

Let $(Z_t)_{t \ge 0}$ an age-dependent process determined by f and G, and $(\zeta_n)_{n \ge 0}$ the process defined by:

$\zeta_n :=$ "the number of individuals in the process $(Z_t)_{t \ge 0}$ belongs to nth generation"

The process $(\zeta_n)_{n \ge 0}$ is the Galton-Watson process with probability generating function $f(s)$.

The probability of extinction $q = P\left(Z_t \to 0\right)$ of the process $(Z_t)\, t \ge 0$, is the non-negative small root of the equation $f(s) = s$.

Definition 4.9 The Malthusian parameter ρ is defined as the root (if exists) of the equation:

$$m \int_0^{\infty} \exp\left(-\rho t\right) dG\left(t\right) = 1.$$

Remark 4.4 If $1 < m < \infty$, then the Malthusian parameter ρ always exists. If $m < 1$, then the Malthusian parameter ρ does not exist (always), however, if it exists, it is necessarily negative.

Theorem 4.8 *If* $(Z_t)_{t \ge 0}$ *is an age-dependent process determined by* f *and* G, *then* $M(t) := E\left(s^{Z_t}\right)$ *satisfies the following equation:*

$$M(t) = [1 - G(t)] + m \int_0^t M(t - u)\, dG(u)$$

Next we will work with the special case in which the distributions of the life length of individuals are independent and identically distributed random variable with exponential distribution of parameter $\alpha > 0$. At the time of the death each individual generates number of random children (all of the same type) according to a offspring distribution $(p_k)_{k \ge 0}$. The process $(Z_t)_{t \ge 0}$, where Z_t denotes the number of individuals present at time t. The process turns out to be Markov chain with continuous-time parameter, this is also called Markov branching process with continuous-time parameter and distribution of offspring $(p_k)_{k \ge 0}$. As in the Galton-Watson process, the distribution of Z_{t+t_0} given that $Z_{t_0} = i$ is equal to the distribution of a sum of i independent and identically distributed random variables of Z_t given that $Z_0 = 1$.

Definition 4.10 Let $(Z_t)_{t\geq 0}$ a Markov branching process with continuous time parameter and set of states S. We define:

$$T_0 := 0$$
$$T_1 := \inf \{t \geq 0 : Z_t \neq Z_0\}$$
$$T_2 := \inf \{t \geq T_1 : Z_t \neq Z_{T_1}\}$$
$$\vdots$$
$$T_n := \inf \{t \geq T_{n-1} : Z_t \neq Z_{T_n}\}$$

This is, T_1 represents the time when the process changes for first time its initial state, T_2 corresponds the time when the process changes for second time its state, and successively. The random variables

$$\tau_n := T_n - T_{n-1}$$

represent the times when the process remains constant in some states, called passage time.

Definition 4.11 The process $(X_n)_{n\geq 0}$ where $X_n := Z_{T_n}$ is a Galton-Watson process with offspring distribution $(p_k)_{k\geq 0}$ called *Skeletons of branching processes* $(Z_t)_{t\geq 0}$.

If $Z_0 = i$ then the first jump in the population size occurs when one of those i initial individuals dies and reproduces. Therefore, the time T_1, in which occurs that first jump, is equal to min $\{V_1, V_2, \ldots, V_i\}$ where, for $k = 1, 2, \ldots, i$, V_k is the life time of k initial individual. therefore,

$$T_1 \overset{d}{=} Exp\,(i\alpha)$$

With the same path if $X_n := Z_{T_n} = i$ then $\tau_n \overset{d}{=} Exp\,(i\alpha)$.

Definition 4.12 Let $(Z_t)_{t\geq 0}$ a Markov branching process with continuous-time. We define:

1. The transition probabilities

$$P_{ij}\,(t) := P\,(Z_t = j \mid Z_0 = i)$$

for all $t \geq 0$.
2. The probabilities generating function of a random variable Z_t with $Z_0 = i$

$$F_i\,(s, t) := E\left(s^{Z_t} \mid Z_0 = i\right)$$

where $s \in [0, 1]$ and $t \geq 0$.

We see that $(Z_t)_{t\geq 0}$ is a homogeneous Markov chain with transition probability

$$P_{ij}(t) = P(Z_{t+s} = j \mid Z_s = i)$$

By using continuous-time Markov chain theory, we have the transition matrices $\mathbf{P}(t) :=$ $\left(P_{ij}(t)\right)_{i,j\in S}$ with $t \geq 0$ have the semi-group property. This is:

$$\mathbf{P}(t+s) = \mathbf{P}(s)\mathbf{P}(t) = \mathbf{P}(t)\mathbf{P}(s)$$

for all $s, t \geq 0$.

Notation If $Z_0 = 1$, then we can simplify the notation with by

$$p_j(t) = P_{1j}(t)$$

and

$$F(s,t) := E\left(s^{Z_t} \mid Z_0 = 1\right)$$

The Markov branching process has an infinitesimal generator Q of the process $(Z_t)_{t\geq 0}$ is equal to:

$$Q = \alpha \begin{pmatrix} 0 & 0 & 0 & 0 & \cdots\cdots \\ p_0 & -1 & p_2 & p_3 & \cdots\cdots \\ 0 & 2p_0 & -2 & 2p_2 & \cdots\cdots \\ 0 & 0 & 3p_0 & -3 & \cdots\cdots \\ \vdots & \vdots & \vdots & \ddots & \cdots\cdots \end{pmatrix}$$

If $p_0 = p_2 = \frac{1}{2}$ and $p_k = 0$ for all $k \neq 0, 2$ then

$$Q = \alpha \begin{pmatrix} 0 & 0 & 0 & 0 & 0 & \cdots \\ \frac{1}{2} & -1 & \frac{1}{2} & 0 & 0 & \cdots \\ 0 & 1 & -2 & 1 & 0 & \cdots \\ 0 & 0 & \frac{3}{2} & -3 & \frac{3}{2} & \cdots \\ \vdots & \vdots & \vdots & \ddots & \cdots\cdots \end{pmatrix}$$

for all $t \geq 0$, the transition matrix $\mathbf{P}(t)$ of the process $(Z_t)_{t\geq 0}$ satisfies the Kolmogorov differential equations. This is, for all $i, j \in \mathbb{N}$ we have:

$$\frac{\partial P_{ij}(t)}{\partial t} = i\alpha p_0 P_{i-1,j}(t) - i\alpha P_{ij}(t) + i\alpha \sum_{k\geq 1}^{i} p_{k+1} P_{i+k,j}(t)$$

$$P_{ij}(0) = \delta_{ij}$$

In particular, If $p_0 = p_2 = \frac{1}{2}$ and $p_k = 0$ for all $k \neq 0, 2$ then

$$\frac{\partial P_{ij}(t)}{\partial t} = i\frac{\alpha}{2}P_{i-1,j}(t) - i\alpha P_{ij}(t) + i\alpha\frac{1}{2}P_{i+1,j}(t)$$
$$P_{ij}(0) = \delta_{ij}.$$

Remark 4.5 Using the continuous-time Markov chain theory, we know that the solution exists $(P_{ij}(t))_{t\geq0}$ of the Kolmogorov equation that satisfies the condition

$$\sum_j P_{ij}(t) \leq 1, \ i = 0, 1, 2, \ldots$$

However, if there are solutions to the equation that satisfy

$$\sum_j P_{ij}(t) < 1, \ i = 0, 1, 2, \ldots$$

then the Kolmogorov equation has infinite solutions.

Since

$$P(Z_t \in \mathbb{N} \mid Z_0 = i) = \sum_{j=0}^{\infty} P(Z_t = j \mid Z_0 = i)$$
$$= \sum_{j=0}^{\infty} P_{ij}(t)$$

then the condition

$$\sum_j P_{ij}(t) < 1$$

implies that with probability greater than zero, and $Z_t = \infty$, the population exploits in a finite interval.

Lemma 4.1 Let $(Z_t)_{t\geq0}$ a Markov branching process with continuous-time. Then

$$F_i(s, t) = (F(s, t))^i$$

for all $s \in [0, 1]$ and $t \geq 0$.

Definition 4.13 Let $(Z_t)_{t\geq0}$ be a Markov branching process with continuous-time parameter with offspring distribution $(p_k)_{k\geq0}$. We define:

$$f(s) := \sum_{k=0}^{\infty} p_k s^k$$

with $s \in [0, 1]$ and

$$u(s) := \alpha \left(f(s) - s \right).$$

A necessary and sufficient condition is given by Harris [5], that no explosions occur in finite time intervals. We now state the following theorem without proof.

Theorem 4.9 *The solution* $\left(P_{ij}(t) \right)_{t \geq 0}$ *of the Kolmogorov equation satisfies the condition*

$$\sum_j P_{ij}(t) = 1$$

if and only if, for all $\varepsilon > 0$ *we have*

$$\int_{1-\varepsilon}^1 \frac{du}{f(u) - u} = \infty.$$

Corollary 4.1 *If* $\mu := f'(1) < \infty$ *then*

$$\sum_j P_{ij}(t) = 1.$$

Proof We have

$$\begin{aligned}
f'(1) &= \lim_{u \to 1} \frac{f(u) - f(1)}{u - 1} \\
&= \lim_{u \to 1} \frac{f(u) - 1}{u - 1}
\end{aligned}$$

then, when $u \to 1$,

$$f(u) - u = \left(f'(1) - 1 \right)(u - 1) + o(u - 1)$$

and then

$$\int_{1-\varepsilon}^1 \frac{du}{f(u) - u} = \infty$$

We will assume $m < \infty$.

Remark 4.6 Since $(Z_t)_{t \geq 0}$ is a Markov chain in continuous-time, with the Kolmogorov forward and backward equations, we can obtain the following results:

$$\frac{\partial F(s, t)}{\partial t} = u(s) \frac{\partial F(s, t)}{\partial s}$$

and

$$\frac{\partial F(s,t)}{\partial t} = u F(s.t)$$

with initial condition

$$F(s,0) = s.$$

Example 4.16 If $p_0 = p_2 = \frac{1}{2}$ and $p_k = 0$ for all $k \neq 0, 2$ then

$$f(s) = \frac{1}{2} + \frac{1}{2}s^2$$

and

$$u(s) = \frac{\alpha}{2}(1-s)^2$$

The probability generating function $F(s,t) = E\left(s^{Z_t} \mid Z_0 = 1\right)$ satisfies the following differential equation

$$\frac{\partial F(s,t)}{\partial t} = u(F(s.t))$$

$$= \frac{\alpha}{2}(1 - F(s,t))^2$$

with the initial condition

$$F(s,0) = s.$$

When we solve the equation, we get

$$F(s,t) = 1 - \frac{1-s}{1 + \frac{\alpha}{2}t(1-s)}.$$

Example 4.17 If the offspring distribution of the process is binomial with parameters $n = 3$ and $p = \frac{3}{4}$ then

$$f(s) = \sum_{k=0}^{3} \binom{3}{k}\left(\frac{3}{4}\right)^k \left(\frac{1}{4}\right)^{3-k} s^k$$

$$= \sum_{k=0}^{5} \binom{3}{k}\left(\frac{3}{4}s\right)^k \left(\frac{1}{4}\right)^{3-k}$$

$$= \left(\frac{3}{4}s + \frac{1}{4}\right)^3$$

and

$$u(s) = \alpha\left(\left(\frac{3}{4}s + \frac{1}{4}\right)^3 - s\right)$$

and we have

$$\frac{\partial F(s,t)}{\partial t} = \alpha \left(\left(\frac{3}{4} F(s,t) + \frac{1}{4} \right)^3 - F(s,t) \right)$$

$$F(s,0) = s.$$

Expected value and variance of the random variable Z_t

In this section, we will assume that $Z_0 = 1$. From the definition of the function $F(s,t)$ we have that

$$m(t) = E(Z_t) = \frac{\partial F(s,t)}{\partial s} \Big|_{s=1}$$

with

$$\frac{\partial F(s,t)}{\partial s} = u(F(s,t))$$

$$F(s,0) = 0$$

we can deduce:

$$\frac{\partial m(t)}{\partial t} = \frac{\partial}{\partial s}(u(F(s,t))) \Big|_{s=1}$$

$$= u'(F(1,t)) m(t)$$

$$= u'(1) m(t)$$

$$= \alpha(\mu - 1) m(t)$$

this is,

$$m(t) = E(Z_0) \exp(\rho t)$$

$$= \exp(\rho t)$$

where $\rho := \alpha(\mu - 1)$. The constant ρ is called Malthus' parameter.

Remark 4.7 1. If $\rho = 0$ then $m(t) = E(Z_0)$ for all $t \geq 0$. We say that the process $(Z_t)_{t \geq 0}$ is critical.
2. If $\rho < 0$ then

$$\lim_{t \to \infty} m(t) = 0$$

In that case the process $(Z_t)_{t \geq 0}$ is called sub-critical.
3. If $\rho > 0$ then

$$\lim_{t \to \infty} m(t) = \infty$$

In that case the process $(Z_t)_{t \geq 0}$ is called super-critical.

On the other hand,

$$Var\,(Z_t) = \frac{\partial^2 F\,(s,t)}{\partial s^2}\,|_{s=1} + m\,(t) - (m\,(t))^2$$

If $v\,(t) := E\left(Z_t^2\right)$ then we have $v\,(t)$ satisfies the equation:

$$
\begin{aligned}
\frac{\partial v\,(t)}{\partial t} &= u''\,(F\,(1,t))\,m^2\,(t) + \rho v\,(t) \\
&= \alpha h''\,(1)\,m^2\,(t) + \rho v\,(t) \\
&= \alpha\left(\sigma^2 - \mu - \mu^2\right)m^2\,(t) + \rho v\,(t) \\
&= \left(\alpha\sigma^2 + \mu\alpha\,(\mu - 1)\right)m^2\,(t) + \rho v\,(t) \\
&= \left(\alpha\sigma^2 + \mu\rho\right)\exp\,(2\rho t) + \rho v\,(t)
\end{aligned}
$$

where $\mu = E\,(Z_0)$ and $\sigma^2 := Var\,(Z_0)$.
therefore:

$$
v\,(t) = \begin{cases} \left(\alpha\sigma^2 + \mu\rho\right)\frac{\exp(2\rho t) - \exp(\rho t) -}{\rho} + \exp\,(\rho t) & \text{if } \rho \neq 0 \\ \alpha\sigma^2 t + 1 & \text{if } \rho = 0 \end{cases}
$$

and

$$
Var\,(Z_t) = \begin{cases} \left(\alpha\sigma^2 + (\mu - 1)\,\rho\right)\frac{\exp(2\rho t) - \exp(\rho t) -}{\rho} & \text{if } \rho \neq 0 \\ \alpha\sigma^2 t & \text{if } \rho = 0 \end{cases}
$$

Definition 4.14 Let $(Z_t)_{t\geq 0}$ be a Markov branching process in continuous-time with dis tribution $(p_k)_{k\geq 0}$ con $Z_0 = 1$. The extinction probability of the process in t is given by:

$$e\,(t) := P\,(Z_t = 0).$$

Remark 4.8 Using the definition of the function $F\,(s,t)$ we have that:

$$e\,(t) = F\,(0,t)$$

and

$$
\frac{de\,(t)}{dt} = \frac{\partial F\,(0,t)}{\partial t} = u\,(F\,(0.t)) = u\,(e\,(t))
$$
$$
e\,(0) = P\,(Z_0 = 0) = 0
$$

If $T_e :=$ "extinction time of the process" then the distribution function of T_e is equal to:

$$P\,(T_e \leq t) = e\,(t)$$

Example 4.18 If $p_0 = p_2 = \frac{1}{2}$ and $p_k = 0$ for all $k \neq 0, 2$ then

$$e(t) = F(0, t)$$
$$= 1 - \frac{1}{1 + \frac{\alpha}{2}t}$$

and

$$E(T_e) = \int_0^\infty P(T_e > t)\,dt$$
$$= \int_0^\infty \frac{1}{1 + \frac{\alpha}{2}t}\,dt$$
$$= \int_0^\infty \frac{2}{2 + \alpha t}\,dt = \infty.$$

Example 4.19 The most familiar of the Markov branching processes is the birth and death process as discussed earlier, in which it is assumed that any individual existing in time t has death probability $\mu(t)\,dt$ in the interval time $(t, t + dt)$ and a probability of $\lambda(t)\,dt$ to disappear and be replaced by two new individuals of the same type in the same time interval. In the special case in which the birth and death rates do not depend t, we obtain

$$p_0 = \frac{\mu}{\lambda + \mu}$$
$$p_2 = \frac{\lambda}{\lambda + \mu}$$
$$p_k = 0, \ k \neq 0, 2$$

with λ and μ constants bigger than 0 and $\alpha = \lambda + \mu$.
   In this case

$$f(s) = p_0 + p_2 s^2$$
$$= \frac{\mu}{\lambda + \mu} + \frac{\lambda}{\lambda + \mu} s^2, \ s \in [0, 1]$$

and

$$u(s) = \lambda s^2 - (\lambda + \mu) s + \mu$$

The forward Kolmogorov equations are given by:

$$\frac{\partial F(s, t)}{\partial t} = \left(\lambda s^2 - (\lambda + \mu) s + \mu\right) \frac{\partial F(s, t)}{\partial s}$$
$$F(s, 0) = s.$$

With an objective to find the solution to the above system, we consider a transformation $x \longmapsto (s(x), t(x))$ such that $0 \longmapsto (s, 0)$.

Let

$$\tilde{F}(x) := F(s(x), t(x))$$
$$= \text{constant}$$
$$= F(s(0), t(0))$$
$$= F(s, 0)$$
$$= s$$

then

$$0 = \frac{d\tilde{F}(x)}{dx}$$
$$= \frac{\partial F}{\partial s}\frac{ds}{dx} + \frac{\partial F}{\partial t}\frac{dt}{dx}$$

From the Kolmogorov equations we obtain:

$$\frac{\partial F}{\partial t} - \left(\lambda s^2 - (\lambda + \mu)s + \mu\right)\frac{\partial F}{\partial s} = \frac{\partial F}{\partial s}\frac{ds}{dx} + \frac{\partial F}{\partial t}\frac{dt}{dx}$$

where

$$\frac{ds}{dx} = -\left(\lambda s^2 - (\lambda + \mu)s + \mu\right)$$
$$\frac{dt}{dx} = 1$$
$$\frac{dF}{dx} = 0$$

with initial conditions

$$s(0) = s$$
$$t(0) = 0$$
$$\tilde{F}(0) = 0$$

Solving the equation

$$\frac{ds}{dx} = -\left(\lambda s^2 - (\lambda + \mu)s + \mu\right)$$

we observe that if $\lambda \neq \mu$ then

$$\lambda s^2 - (\lambda + \mu)s + \mu = \lambda(s - 1)(s - \rho)$$

with

$$\rho := \frac{\mu}{\lambda}$$

therefore:

$$\frac{ds}{(s-1)(s-\rho)} = -\lambda dx$$

that is,

$$\frac{1}{1-\rho}\left(\frac{1}{s-1} - \frac{1}{s-\rho}\right) ds = \lambda dx$$

$$\left(\frac{1}{s-1} - \frac{1}{s-\rho}\right) = -(\lambda - \mu) dx$$

then

$$\int_{s(0)}^{s(x)}\left(\frac{1}{s-1} - \frac{1}{s-\rho}\right) ds = -(\lambda - \mu) x$$

in other words,

$$\ln\left(\frac{s(x)-\rho}{s(x)-1}\right) - \ln\left(\frac{s-\rho}{s-1}\right) = (\lambda - \mu) x$$

$$\frac{s(x)-\rho}{s(x)-1} = \left(\frac{s-\rho}{s-1}\right)\exp\left(-(\lambda - \mu) x\right)$$

therefore,

$$s(x) = \frac{1-\rho}{1 - \left(\frac{s-1}{s-\rho}\right)\exp\left(-(\lambda - \mu) x\right)}$$

As $t(x) = x$, $\tilde{F}(x) = F(s, 0) = 0$ then for $\lambda \neq \mu$ is obtained:

$$F(s,t) = \frac{\mu(s-1) - (\lambda s - \mu)\exp\left(-(\lambda - \mu) t\right)}{\lambda(s-1) - (\lambda s - \mu)\exp\left(-(\lambda - \mu) t\right)}$$

then

$$e(t) = F(0, t)$$
$$= \frac{-\mu + \mu\exp\left(-(\lambda - \mu) t\right)}{-\lambda + \mu\exp\left(-(\lambda - \mu) t\right)}$$
$$= \frac{-\rho(1 - \exp\left(-(\lambda - \mu) t\right))}{1 + \rho\exp\left(-(\lambda - \mu) t\right)}$$

and therefore

$$q := \lim_{t\to\infty} q(t) = \begin{cases} 1 \text{ if } \mu > \lambda \\ \rho \text{ if } \mu < \lambda \end{cases}$$

For $\lambda = \mu$ we can obtain.

$$\frac{ds}{dx} = -\left(\lambda s^2 - (\lambda + \mu) s + \mu\right)$$
$$= -\left(\lambda s^2 + 2\lambda s + \lambda\right)$$
$$= -\lambda (s - 1)^2$$

this is,

$$\frac{ds}{(s - 1)^2} = -\lambda dx$$

where,

$$\int_{s(0)}^{s(x)} \frac{ds}{(s - 1)^2} = -\lambda x$$

therefore,

$$\frac{1}{s - 1} - \frac{1}{s(x) - 1} = -\lambda x$$
$$s = 1 + \frac{s(x) - 1}{1 - \lambda x (s(x) - 1)}$$

As $t(x) = x$, $\tilde{F}(x) = F(s, 0) = 0$ then for $\lambda = \mu$ is obtained:

$$F(s, t) = \frac{s + (1 - s) \lambda t}{1 + (1 - s) \lambda t}$$

and

$$e(t) = F(0, t) = \frac{\lambda t}{1 + \lambda t}$$

and

$$q := \lim_{t \to \infty} q(t) = 1.$$

Example 4.20 The Yule process is a particular case of birth and death. in which it is assumed $\lambda(t) = \lambda > 0$ and $\mu(t) = \mu = 0$. In this case:

$$F(s, t) = \frac{\lambda s \exp(-\lambda t)}{\lambda(s - 1) - (\lambda s) \exp(-\lambda t)}$$
$$= \frac{s \exp(-\lambda t)}{s(1 - \exp(-\lambda t) - 1)}$$

and

$$e(t) = F(0, t) = 0.$$

Example 4.21 A possible modification of the process consists of imposing the condition that only the individuals whose lifespan l is greater than a fixed time $T > 0$ reproduce according to the given distribution of offspring $(p_k)_{k \geq 0}$ while those individuals whose age is less than T dies without producing offspring [6]. In this case, the generating function of Z_t satisfies the following equation:

$$F(s,t) = \begin{cases} s(1-G(t)) + G(t) & \text{if } t \leq T \\ s(1-G(t)) + G(T) + \int_T^t f(F(s, t-u)) \, dG(u) & \text{if } t > T \end{cases}$$

Furthermore, we have that $M(t)$ satisfies the equation

$$M(t) = \begin{cases} 1 - G(t) & \text{if } t \leq T \\ [1-G(t)] + m\int_T^t M(t-u)\, dG(u) & \text{if } t > T \end{cases}$$

where $m = f'(1)$. If $G(t) = 1 - \exp(-\alpha t)$, $t \geq 0$, then

$$M(t) = \begin{cases} \exp(-\alpha t) & \text{if } t \leq T \\ \exp(-\alpha t) + m\int_T^t M(t-u)\,\alpha \exp(-\alpha u)\, du & \text{if } t > T \end{cases}$$

This above equation can be solved by the iteration method, and we get

$$M(t) = \exp(-\alpha t) \sum_{r=0}^{n-1} \left[\frac{\alpha m (t - rT)^r}{r!} \right], \quad (n-1)T < t \leq nT$$

The process $(\zeta_n)_{n\geq 0}$ where

$\zeta_n :=$ "the number of individuals in the process $(Z_t)_{t\geq 0}$ belonging to the nth generation"

is a Galton-Watson process and its offspring probability generating function is given by:

$$h(s) = G(T) + [1 - G(T)] f(s)$$

where f is the offspring probability generating function of the process $(Z_t)_{t\geq 0}$.
    Therefore, the extinction probability of the process with probability 1, if and only if,

$$m(1 - G(T)) \leq 1$$

and the probability of extinction of the process is the smallest root in $[0, 1]$ of the equation

$$G(T) + [1 - G(T)] f(s) = s$$

If the offspring distribution of the process $(Z_t)_{t\geq 0}$ is given by

$$p_k = bc^{k-1}, \quad k = 1, 2, \ldots$$

and

$$p_0 = \frac{1 - b - c}{1 - c}$$

where $0 < b, c < 1$ then

$$f(s) = \frac{1 - b(1 - s)}{(1 - c)(1 - cs)}$$

If $T > 0$ then the probability of extinction of the process $(Z_t)_{t \geq 0}$ is given by:

$$\frac{1}{c} - b \left[\frac{1 - G(T)}{1 - c} \right]$$

and if $T = 0$, then the probability of extinction of the process is equal to:

$$\frac{1}{c} - b \left[\frac{1}{1 - c} \right].$$

4.4   Branching Process Model for COVID-19

Coronaviruses are a family of viruses that can cause mild respiratory illnesses to cases of Severe Acute Respiratory Infection, such as SARS-CoV-2, the virus that causes COVID-19. We explain the methodology for the description of the evolution of the SARS-CoV-2 virus in the city of Bogota, Colombia. In this section we follow our recent work [7][1] and consider a type-two Galton-Watson process with individuals, namely reported infected and unreported infected (asymptomatic) individuals, using the daily statistics of COVID-19 cases reported by the Bogota health secretary.

Let $(Z_n)_n$ be the multi-type Galton-Watson process with $Z_n = \{Z_1(n), Z_2(n)\}$ where $Z_1(n)$ and $Z_2(n)$ correspond, respectively, to the number of types T_1, that is, infected, asymptomatic and unregistered individuals, and to the number of individuals infected and registered, called individuals type T_2, present at time n. It is observed that $Z_1(n)$ is the total number of individuals of type T_1, on day n, infected by the infected individuals present on the day $(n - 1)$ and that $Z_2(n)$ is the total number of registered individuals with COVID-19 present on day n. The process is assumed to begin with $Z_1(0)$ infected individuals. Let

$\xi_1^{(1)}(n; j)$: Number of infected individuals type T_1 on day n, infected by the jth individual on day $(n - 1)$ where $j = 1, 2, 3 \ldots, Z_1(n - 1)$.

$\xi_2^{(1)}(n; j)$: Number of registered infected individuals with COVID-19 of type T_2 on day n, infected by the jth infected individual of $(n - 1)$th day with $j = 1, 2, 3 \ldots, Z_1(n - 1)$.

The vector of offspring ξ_1 of individuals of type T_1 is given by $\xi_1 = \left(\xi_1^{(1)}, \xi_2^{(1)} \right)$, where the components indicate the total number of offspring of type T_1 and type T_2, of the initially infected individuals. The joint probabilities generating function of offspring of $\xi_1^{(1)}$ and $\xi_2^{(1)}$ is given by [8]:

---

[1] We thank Editor of the Journal *Ciencia en Desarollo* for the permission to reprint figures and related results.

$$h_1(s_1, s_2) = \mathbf{E}\left(s_1^{\xi_1^{(1)}} s_2^{\xi_2^{(1)}}\right) = p_0 + \sum_{j=1}^{k} p_j s_1^j + q s_2$$

$q = 1 - \sum_{j=0}^{k} p_j$, $h_1(1, 1) = 1$, where $|s_1| \le 1$, $|s_2| \le 1$

where p_0 is the probability that individuals of type T_1 exit the place of propagation (for example, they migrate to another country), p_j is the probability of producing j new contaminated of type T_1 and q is the probability that an individual of type T_1 is confirmed sick or dead. Since individuals of type T_2 do not reproduce as they are isolated or quarantined, then the corresponding probability generating function is $h_2(s_1, s_2) \equiv 1$.

The probability generating functions of $\xi_1^{(1)}$ and $\xi_2^{(1)}$ are given by:

$$\mathbf{E}\left(s_1^{\xi_1^{(1)}}\right) = h_1(s_1, 1) = p_0 + \sum_{j=1}^{k} p_j s_1^j + q = 1 - \sum_{j=1}^{k} p_j\left(1 - s_1^j\right)$$

$$\mathbf{E}\left(s_2^{\xi_2^{(1)}}\right) = h_1(1, s_2) = 1 - q + q s_2$$

Assume $Z_1(0) > 0$ and $Z_2(0) = 0$ so for $n = 1, 2, \ldots$

$$Z_1(n) = \sum_{j=1}^{Z_1(n-1)} \xi_1^{(1)}(n; j) \qquad Z_2(n) = \sum_{j=1}^{Z_1(n-1)} \xi_2^{(1)}(n; j),$$

where the vectors $\left\{\left(\xi_1^{(1)}(n; j), \xi_2^{(1)}(n; j)\right)\right\}$ are independent and identically distributed (i.i.d) as $\left(\xi_1^{(1)}, \xi_2^{(1)}\right)$.

Since $\mathbf{P}\left\{\xi_2^{(1)}(n; j) = 0\right\} = 1 - q$ and $\mathbf{P}\left\{\xi_2^{(1)}(n; j) = 1\right\} = q$, then the process $Z_2(n)$ has a binomial distribution of parameters $Z_1(n - 1)$ and q. That is,

$$\mathbf{P}\{Z_2(n) = i \mid Z_1(n - 1) = l\} = \binom{l}{i} q^i(1 - q)^{l-i}, i = 0, 1, \ldots, l; l = 0, 1, 2, \ldots$$

That is, q can be interpreted as the proportion of infected individuals registered within the group of infected individuals on day $(n - 1)$.

To determine the average number of individuals infected but not registered in the sample, use will be made of the probability generating functions of $Z_1(n)$ and $Z_2(n)$, following the usual procedure according to to [8].

Let

$$h_0(s) = \mathbf{E}s^{Z_1(0)}, \quad F_1(n; s) = \mathbf{E}\left(s^{Z_1(n)}\right), \quad F_2(n; s) = \mathbf{E}\left(s^{Z_2(n)}\right).$$

$$h^*(s) = h_1(s, 1) = q + p_0 + \sum_{j=1}^{k} p_j s^j \quad and$$

$$\tilde{h}(s) = h_1(1, s) = 1 - q + q s$$

Let $h_n(s_1, s_2)$ the joint probability generating function of $\xi_1^{(1)}(n; j)$ and $\xi_2^{(1)}(n; j)$, $h_n^* :=$ $h_n(s, 1)$ and $\tilde{h}_n = h_n(1, s)$. Is easy to verify that

$$F_1(n; s) = \mathbf{E}\left(s^{Z_1(n)}\right) = F_1\left(n-1; h^*(s)\right) = F_1\left(0; h_n^*(s)\right) = h_0\left(h^*\left(h^*\left(\dots\left(h^*(s)\right)\dots\right)\right)\right)$$
$$F_2(n; s) = \mathbf{E}\left(s^{Z_2(n)}\right) = F_1\left(n-1; \tilde{h}(s)\right) = F_1\left(0; \tilde{h}_n(s)\right) = h_0\left(\tilde{h}\left(\tilde{h}\left(\dots\left(\tilde{h}(s)\right)\dots\right)\right)\right)$$

It is then found that the average number of individuals infected by an unreported contaminated individual (c.i.) equals

$$m = \frac{d}{ds}h^*(s)\Big|_{s=1}$$

and that the average number of individuals infected, not reported, by an unreported c.i. is given by

$$\mathbf{E}\xi_1^{(1)} = \sum_{j=1}^{k} jp_j, k = 1, 2, \dots,$$

It is observed that

$$\frac{d}{ds}\tilde{h}(s)\Big|_{s=1} = \mathbf{E}\xi_2^{(1)} = q$$

is the average value of individuals infected by a registered c.i.

From the equations given above it is obtained that:

$$M_1(n) = \mathbf{E}Z_1(n) = m_0 m^n, n = 0, 1, 2, \dots \quad and$$

$$M_2(n) = \mathbf{E}Z_2(n) = q\mathbf{E}Z_1(n-1) = qm_0 m^{n-1}, n = 1, 2, \dots$$

where, $m_0 = 1$ and $\mathbf{E}(Z_2(0)) = 0$.

Parameters estimation

We notice that $\frac{\mathbf{E}Z_2(n+1)}{\mathbf{E}Z_2(n)} = m$. So (see [9]),

$$\widehat{m}_n = \frac{Z_2(n+1)}{Z_2(n)}$$

is an empirical estimator of the parameter m, which is also a strongly consistent estimator (see [10]).

The maximum likelihood estimator called, **Harris estimator**[10], is given by:

$$\tilde{m}_n = \sum_{i=2}^{n+1} Z_2(i) \Big/ \sum_{j=1}^{n} Z_2(j), n = 1, 2, \dots$$

Finally, we will consider the **Crump-Hove estimator** [11] which is given by:

$$\bar{m}_{n,N} = \sum_{i=n+1}^{n+N} Z_2(i) / \sum_{j=n}^{n+N-1} Z_2(j), n = 1, 2, \ldots; N = 1, 2, \ldots$$

Thus, when estimating the value of m, the expected value of unreported individuals in the infected population with COVID-19 can be predicted with $Z_1(0) = 1$, and $M_1(n) = \mathbf{E} Z_1(n)$. These can be approximated by $\widehat{m}n^n$, $\widetilde{m}_n^n$, or $\bar{m}_{n,N}^n$ respectively. We obtain the following three estimators of $M_1(n)$:

$$\widehat{M}_1(n) = \widehat{m}_n^n,$$

$$\widetilde{M}_1(n) = \widetilde{m}_n^n \text{ and}$$

$$\bar{M}_1(n) = \bar{m}_{n,N}^n$$

Therefore, if we have the observations $(Z_2(1), Z_2(2), \ldots, Z_2(n))$ in the first n days, the average value of infected individuals in the next k days can be predicted through the following relationships;

$$\widehat{M}_1(n + k) = \widehat{m}_n^{n+k}, \ \widetilde{M}_1(n + k) = \widetilde{m}_n^{n+k} \text{ and } \bar{M}_1(n + k) = \bar{m}_{n,N}^{n+k}, k = 1, 2, \ldots$$

Proportion of registered infected individuals (T_2) in the population.

According to [11], the estimated proportion of individuals infected and registered on the nth day is given by:

$$\widehat{\alpha}(n) = Z_2(n) / \left\{ Z_2(n) + \widehat{M}_1(n) \right\},$$

$$\widetilde{\alpha}(n) = Z_2(n) / \left\{ Z_2(n) + \widetilde{M}_1(n) \right\} \quad and$$

$$\bar{\alpha}(n) = Z_2(n) / \left\{ Z_2(n) + \bar{M}_1(n) \right\}.$$

Computational implementation of the model

For the computational part of the model, the reports are registered by the page of the open data portal of the Colombian State [12] of the city of Bogotá. The data collected is public information available to citizens under an open license and without legal restrictions in accordance with Colombian Law on Transparency and Access to National Public Information in Colombia. Finally, for processing and analyzing these records, Python version 3.7 software was implemented and is given in the supplied database. The data on the nth day correspond to daily individuals reported by COVID-19 in Bogota; that is, they are the values $Z_2(n)$. The data of the *new* infected individuals, $Z_1(n)$, are unknown as well as the initial value of this process $m_0 = E Z_1(0)$.

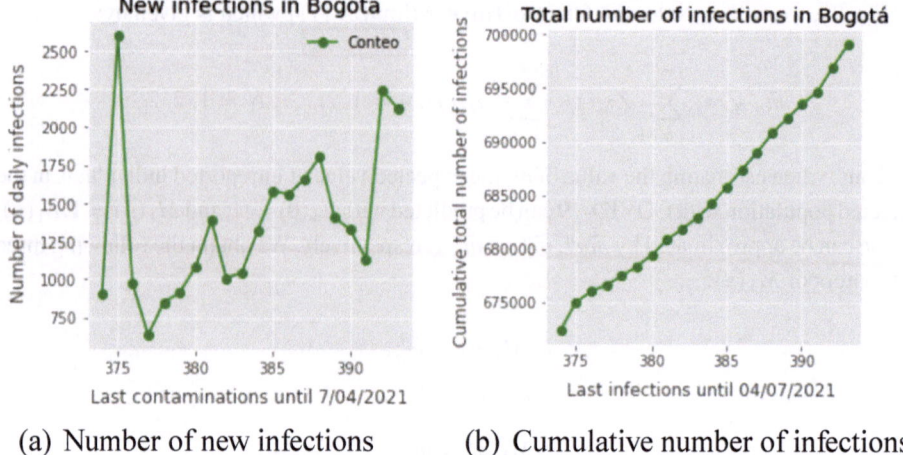

(a) Number of new infections   (b) Cumulative number of infections

Fig. 4.2 Graphs showing the number of new infections (**a**) and the cumulative number of infections (**b**) (see [7])

Figure 4.2a and b respectively, shows the graphs of the new cases registered daily and the total accumulated during the last 20 days before April 7, 2021 (date on which the analysis is made for this work) in the city of Bogota.

The contagion parameter m for new infections is estimated for each of the three previously proposed estimators. The behavior of this value can be seen in Fig. 4.3a. This value indicates

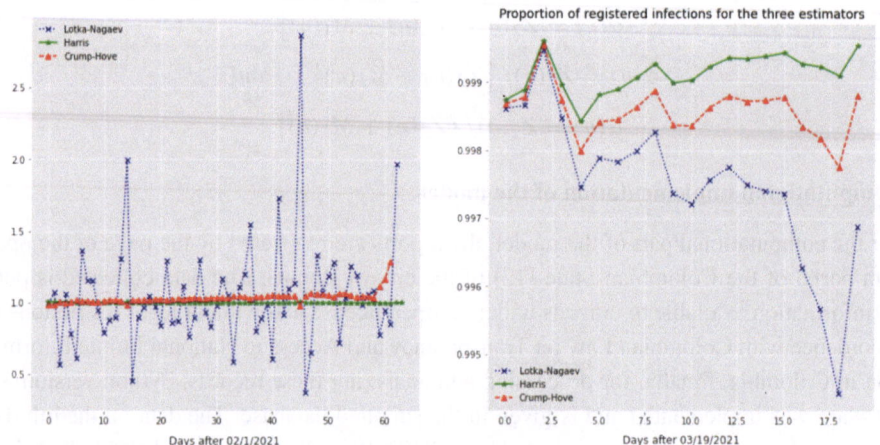

(a) Average contamination parameter   (b) Proportion of registered individuals
m of new registered case

Fig. 4.3 Graphs showing the average contamination parameter m of new registered case (**a**) and the proportion of registered individuals (**b**) (see [7])

the expected number of new infections caused by an infected individual. These values correspond to the last 60 days that occurred after February 1, 2021. The Harris and Crump-Hove estimators show a relatively more stable behavior during the period analyzed, that is, these values do not vary much as the years go by days. Figure 4.3b shows the corresponding values for the proportion of registered cases $Z_2(n)$ with COVID-19 among the total infected population (individuals of type T_1 or T_2) for each estimator, this value is very close to 1, this means that close to 99% of the recorded infections correspond to people who tested positive for COVID-19 and these are individuals of type T_2 (symptomatic people).

Table 4.1 represents the estimation value of the reproduction parameter from the virus by an i.c. on the day $Z_2(1)$, data corresponding to March 19, 2021. This value was obtained based on each estimator and is one of the 20 values of the process $Z_2(1), Z_2(2), \cdot, Z_2(20)$, being this the first of the 20 most recent cases in this study.

The expected value of the number of infections by a c.i. for unregistered individuals can be seen in Fig. 4.3b along with a forecast for five days from April 8 to 12 (Table 4.2) after day 20 (last day) for each one of the three proposed estimators.

Table 4.1 Values of m for each estimator

Estimator	Estimation day $n = 1$
$\widehat{M}_1(1)$	1,10
$\widetilde{M}_1(1)$	1,00
$\overline{M}_1(1)$	1,05

Table 4.2 Forecast for 5 days

Day	Lotka-Nagaev	Harris	Crump-Hove
1	7,021	1,024	2,681
2	7,739	1,025	2,816
3	8,531	1,026	2,958
4	9,405	1,027	3,108
5	10,367	1,028	3,256

References

1. Athreya, K. B., Ney, P. E., & Ney, P. (2004). *Branching processes*. Courier Corporation.
2. Rœlly, S. (2017). Algunas propiedades basicas de procesos de ramificacion. In *Stochastic processes with applications in the natural sciences: international workshop at Universidad de los Andes, Bogotá, Colombia* (Vol. 4, p. 29). Universitätsverlag Potsdam.
3. Kraus, S. D., Kenney, R. D., Mayo, C. A., McLellan, W. A., Moore, M. J., & Nowacek, D. P. (2016). Recent scientific publications cast doubt on North Atlantic right whale future. *Frontiers in Marine Science*, 137.
4. Caswell, H., Fujiwara, M., & Brault, S. (1999). Declining survival probability threatens the North Atlantic right whale. *Proceedings of the National Academy of Sciences, 96*(6), 3308–3313.
5. Harris, T. E., et al. (1963). *The theory of branching processes* (Vol. 6). Berlin: Springer.
6. Parthasarathy, P. (1979). On a modified Markov branching process. *Journal of Mathematical Biology, 7*, 95–97.
7. Plata-Bello, G., Blanco, L., & Arunachalam, V. (2022). Procesos de ramificación para modelar el sars-cov-2 en la ciudad de bogotá. *Ciencia en Desarrollo, 13*(2).
8. Yanev, N. M., Stoimenova, V. K., & Atanasov, D. V. (2020). Stochastic modeling and estimation of covid-19 population dynamics. arXiv:2004.00941.
9. Dion, J. -P. (1975). Estimation of the variance of a branching process. *The Annals of Statistics*, 1183–1187.
10. Maaouia, F., Touati, A., et al. (2005). Identification of multitype branching processes. *Annals of Statistics, 33*(6), 2655–2694.
11. Yanev, N. M., Stoimenova, V. K., & Atanasov, D. V. (2020). Branching stochastic processes as models of covid-19 epidemic development. arXiv:2004.14838.
12. GOVCO. (2020). Datos covid-19, https://www.datos.gov.co/Salud-y-Protecci-n-Social/Casos-positivos-de-COVID-19-en-Colombia/gt2j-8ykr/data.

Hidden Markov Model

5

Hidden Markov Model (HMM) was introduced by American mathematician Leonard Baum (1931–2017) in the late 1960s in his work titled *An Inequality and Associated Maximization Technique in Statistical Estimation for Probabilistic Functions of Markov Processes* [1]. Later, with fellow American mathematician Richard Welch (1927-), he developed the so-called Baum-Welch algorithm that allows automatic speech recognition. At the Symposium on applications of hidden Markov models held at Princeton in 1980, the American mathematician Lee P. Neuwirth (1933-) introduced the term Hidden Markov Models instead of *Probabilistic Functions of Markov Processes* [2]. In 1989, the electrical engineer Lawrence R. Rabiner (1943-) published a tutorial on HMM describing Baum's theory applied to speech recognition, an obligatory reference in studying this model [3]. Hidden Markov models have found a wide range of applications. In various areas of applications, such as the analysis of image [4], medicine [5], and ecology [6, 7].

In the 1940s, following the rapid development of computer science, it became necessary to build algorithms that allow for solving real-life problems quickly and efficiently. The situation that researchers frequently face when building models is that the observations obtained are incomplete, either due to the impossibility of physical or by the presence of noise in measurements, for example, when they want to study the movements and behaviors of some animals such as whales, sharks, and leopards in their natural environment, it is not possible to access information directly [7, 8]. However, the measurement of the distances traveled by the animals seeks to infer if they were searching for food, migrating, or resting. In other words, from a sequence of observations, it is desired to determine the real states of the system. Hidden Markov models have been useful in solving this and similar problems.

© The Author(s), under exclusive license to Springer Nature Switzerland AG 2023
L. Blanco-Castañeda and V. Arunachalam, *Applied Stochastic Modeling*, Synthesis
Lectures on Mathematics & Statistics, https://doi.org/10.1007/978-3-031-31282-3_5

5.1   Hidden Markov Chain

The idea of HMM is to see a system we cannot access or observe directly and is a doubly stochastic system. It presents randomness in the observations and the sequence of associated unobserved (hidden) states that generate them. Hidden Markov models arise from the need to study the problem to characterize stochastic processes in which the observations obtained are incomplete either due to physical impossibilities or due to the presence of noise in the measurements, and it is here that new thinking where remarks follow a pattern of hidden behavior and, therefore, the absence of some remarks is irrelevant. Hidden Markov chains have multiple applications: they constitute, for example, a very efficient technique for the problem of automatic speech recognition, which consists of finding the most likely sequence given in a locution.

A hidden Markov chain $\{X_n; n \geq 1\}$ with a set of finite states $S := \{s_1, s_2, \ldots, s_N\}$ is a homogeneous Markov chain that is not directly observable. This model describes the observable events which depend on external factors. Observable events are represented as symbols of a finite alphabet, and the hidden factors involved in the observations constitute the set of states of the chain. That is, the Hidden Markov chain is fully characterized by the following:

$$\lambda = (S, O, A, B, \pi)$$

where

1. $S := \{s_1, s_2, \ldots, s_N\}$ is a finite set representing the possible (hidden) states in the chain.
2. $O := \{O_1, O_2, \ldots, O_M\}$ is the set of possible observations where a each observation is corresponds a symbol v_i of a finite alphabet.
3. $A := (a_{ij})_{N \times N}$ is the matrix of transition between the states of the chain. Namely,

$$a_{ij} := P\left(X_{n+1} = s_j \mid X_n = s_i\right)$$

4. $B := (b_{ij}(O_t))_{N \times M}$ is the transition matrix between the observations and the states of the chain. That is,

$$b_{ij}(O_n) := P\left(O_n = v_j \mid X_n = s_i\right)$$

5. $\pi := (\pi_i)_{n \times 1}$ is the initial distribution of the chain, that is

$$\pi_i := P(X_1 = s_i)$$

Example 5.1 We consider an example [9] in which it is desired to determine in the year 2779, using the climate of the year 2023, either hot (H) or rainy (R), based on the daily observation, when a person who writes down the number of ice creams 1, 2 or 3, daily he ate every day. We have the following sequences of observations and hidden states

$$R\ R\ R\ H\ H\ H\ H\ R\ H\ R$$
$$1\ 1\ 2\ 3\ 2\ 1\ 3\ 2\ 3\ 1$$

$$H\ R\ R\ H\ H\ R\ R\ R\ H\ H$$
$$2\ 2\ 1\ 3\ 3\ 2\ 2\ 1\ 1\ 2$$

and

$$H\ H\ H\ R\ H\ R\ R\ R\ R\ H$$
$$3\ 3\ 2\ 2\ 2\ 1\ 1\ 1\ 3\ 2$$

Let

$$X_t := \text{"the climate of the day } t\text{"}, t = 1, 2, \ldots$$

We consider the following estimates of the initial distribution π, the transition matrix between hidden states A and the matrix of emissions B:

$$\pi_i : = P(X_1 = s_i)$$
$$= \frac{L_{s_i}}{\sum_{s_k} L_{s_k}}$$

where $L_{s_k} :=$ "number of times the initial state is s_k".

In this case $\pi_1 = P(X_1 = H) = \frac{2}{3}$ and $\pi_2 = P(X_1 = R) = \frac{1}{3}$

$$A = (a_{ij})$$

with

$$a_{ij} = P(X_t = s_j \mid X_{t-1} = s_i)$$
$$= \frac{N_{s_i s_j}}{\sum_k N_{s_i s_k}}$$

where

$$N_{s_i s_j} = \text{"the number of transitions of } s_i \text{ a } s_j\text{"}$$

In this case

$$A = \begin{pmatrix} \frac{8}{14} & \frac{6}{14} \\ \frac{6}{13} & \frac{7}{13} \end{pmatrix}$$

and

$$b_i(v_j) = P(O_t = v_j \mid X_t = s_i)$$
$$= \frac{M_{s_i v_j}}{\sum_k M_{s_i v_k}}$$

We have

$M_{s_i v_j}$ = "number of times the hidden state is s_i and it is observed v_k(in the same position)"

In this case

$$B = \begin{pmatrix} \frac{4}{5} & \frac{1}{2} & \frac{1}{8} \\ \frac{1}{5} & \frac{1}{2} & \frac{7}{8} \end{pmatrix}$$

We now state the three significant problems of interest that need to be solved so that the model to be helpful in the application.

1. Evaluation problem: given a hidden Markov chain characterized by $\lambda = (S, O, A, B, \pi)$ and a sequence of observations

$$v_1 v_2 \dots v_T$$

   it seeks to find the model that generates the observed sequence.
2. Decomposition problem: given a hidden Markov chain characterized by $\lambda = (S, O, A, B, \pi)$ and a sequence of observations

$$v_1 v_2 \dots v_T$$

   it seeks to determine the sequence of hidden states that maximize the probability of observing the given sequence.
3. Training problem: given a sequence of observations $v_1 v_2 \dots v_T$ and the general structure of the chain, that is, the sets of visible states and hidden states, we seek to determine the parameters A, B and π of the model that maximize the probability of observing that sequence. In other words, it seeks to determine how the model parameters should be adjusted to maximize the probability that the observation was generated by the model.

The solutions to each of the problems described above are presented below:

Solution of the evaluation problem

Let $\lambda = (S, O, A, B, \pi)$ be the given model and $\mathcal{O} = v_1 v_2 \dots v_T$ a sequence of observations. we want to calculate $P(\mathcal{O} \mid \lambda)$. Suppose $S = \{s_1, s_2, \dots, s_T\}$ is a sequence of states of the chain. By the definition of b_{ij} we have that

$$P(\mathcal{O} \mid S, \lambda) = P(O_1 = v_1 \mid X_1 = s_1) P(O_2 = v_2 \mid X_2 = s_2) \cdots P(O_T = v_T \mid X_T = s_T)$$

and by the definition of π and A we have:

$$P(S \mid \lambda) = P(X_1 = s_1, X_2 = s_2, \dots, X_T = s_T)$$
$$= P(X_1 = s_1) P(X_2 = s_2 \mid X_1 = s_1) \cdots P(X_T = s_T \mid X_{T-1} = s_{T-1})$$

Therefore:

$$P\left(\mathcal{O} \mid \lambda\right) = \sum_{S} P\left(\mathcal{O}, S \mid \lambda\right)$$

$$= \sum_{S} P\left(\mathcal{O} \mid S, \lambda\right) P\left(S \mid \lambda\right)$$

The direct computation of the above expression requires approximately $2nN^n$ multiplications. The following dynamic programming algorithm, called the forward recursion algorithm, allows us to reduce the number of multiplications required to nN^2.

Forward algorithm

For $t = 1, 2, \ldots, T$ and $i = 1, 2, \ldots, N$ we define

$$\alpha_t\left(i\right) := P\left(O_1 = v_1, O_2 = v_2, \ldots, O_t = v_t, X_t = s_i\right)$$
$$= P\left(O_1 = v_1, O_2 = v_2, \ldots, O_t = v_t, X_t = s_i \mid \lambda\right)$$

Here, $\alpha_t\left(i\right)$ is the probability that at time t the chain is in state s_i and that the sequence of observations obtained up to time t is $v_1, v_2, \ldots, v_t$. The probability $\alpha_t\left(i\right)$ is called probability forward. Then:

$$P\left(O_1 = v_1, O_2 = v_2, \ldots, O_t = v_t, X_t = s_i \mid \lambda\right) =$$

$$\sum_{j=1}^{N} P\left(O_1 = v_1, O_2 = v_2, \ldots, O_t = v_t, X_{t-1} = s_j, X_t = s_i \mid \lambda\right)$$

and since the distribution of the observation O_t depends only on the state of X_t and not on any state or any previous observation, then it follows that

$$P\left(O_1 = v_1, O_2 = v_2, \ldots, O_t = v_t, X_{t-1} = s_j, X_t = s_i \mid \lambda\right)$$
$$= P\left(O_1 = v_1, O_2 = v_2, \ldots, O_t = v_t, X_{t-1} = s_j, X_t = s_i\right)$$
$$= P\left(O_1 = v_1, \ldots, O_{t-1} = v_{t-1}, O_t = v_t, X_{t-1} = s_j, X_t = s_i\right)$$
$$= P\left(O_t = v_t, X_t = s_i \mid O_1 = v_1, \ldots, O_{t-1} = v_{t-1}, X_{t-1} = s_j\right)$$
$$\times P\left(O_1 = v_1, \ldots, O_{t-1} = v_{t-1}, X_{t-1} = s_j\right)$$
$$= P\left(O_t = v_t \mid X_t = s_i, O_1 = v_1, \ldots, O_{t-1} = v_{t-1}, X_{t-1} = s_j\right)$$
$$\times P\left(X_t = s_i \mid O_1 = v_1, \ldots, O_{t-1} = v_{t-1}, X_{t-1} = s_j\right)$$
$$\times P\left(O_1 = v_1, \ldots, O_{t-1} = v_{t-1}, X_{t-1} = s_j\right)$$
$$= P\left(O_t = v_t \mid X_t = s_i\right) P\left(X_t = s_i \mid X_{t-1} = s_j\right)$$
$$P\left(O_1 = v_1, \ldots, O_{t-1} = v_{t-1}, X_{t-1} = s_j\right)$$

Therefore, for each $i = 1, 2, \ldots, N$

$$
\alpha_t (i) =
\begin{cases}
\pi_i b_{1i} (O_1) & \text{if } \quad t = 1 \\[2ex]
\left[\displaystyle\sum_{j=1}^{N} \alpha_{t-1} (j) a_{ji} \right] b_{it} (O_t) & \text{if } t = 2, 3, \ldots, T
\end{cases}
$$

Consequently,

$$
P (\mathcal{O} \mid \lambda) = P (O_1 = v_1, O_2 = v_2, \ldots, O_T = v_T)
$$

$$
= \sum_{i=1}^{N} \alpha_T (i)
$$

Algorithm 1: Forward algorithm.

Initialization:

$\alpha_j(1) = \pi_i b_{1i} (O_1) \, \forall i = 1, \ldots, N$

Iteration Step:

for $t = 2, \ldots, T$ **do**

    for $j=1, \ldots, n$ **do**

$$
\alpha_t (i) = \left[\sum_{j=1}^{N} \alpha_{t-1} (j) a_{ji} \right] b_{it} (O_t)
$$

    end

end

Output:

$P (O_1 = v_1, O_2 = v_2, \ldots, O_T = v_T) = \displaystyle\sum_{i=1}^{N} \alpha_T (i)$

Backward Algorithm

The backward algorithm is similar to the forward algorithm except that now it starts at the end time T and work backwards until the start time $t = 1$. is reached.

For $t = 1, 2, \ldots, T$ and for $i = 1, 2, \ldots, N$ $\beta_t (i)$ is defined as follows:

$$
\beta_t (i) = P (O_t = v_t, O_{t+1} = v_{t+1}, \ldots, O_T = v_T \mid X_t = s_i, \lambda)
$$

The probability $\beta t (i)$ can be computed recursively as follows:

$$
\beta_t (i) =
\begin{cases}
1 & \text{if } \quad t = T \\[2ex]
\displaystyle\sum_{j=1}^{N} a_{ij} b_{jt} (O_t) \beta_t (j) & \text{if } t = T - 1, T - 2, \ldots, 1
\end{cases}
$$

Therefore

$$P\left(\mathcal{O} \mid \lambda\right) = P\left(O_1 = v_1, O_2 = v_2, \ldots, O_T = v_T\right)$$

$$= \sum_{i=1}^{N} \beta_1\left(i\right).$$

Algorithm 2: Backward algorithm

Initialization:

$$\beta_T\left(i\right) = P\left(O_T = v_T \mid X_T = s_i\right), \quad for \ 1 \leq i \leq N$$

Iteration Step:

for $t = T\text{-}1, t\text{-}2 \ldots, 1$ **do**

$$\beta_j(T) = \sum_{j=1}^{N} a_{ij} b_{jt}\left(O_t\right) \beta_t\left(j\right)$$

end

Output:

$$P\left(O_1 = v_1, O_2 = v_2, \ldots, O_T = v_T\right) = \sum_{i=1}^{N} \beta_1\left(i\right)$$

Example 5.2 Suppose that two current coins m_1 and m_2 are tossed and that an observer only sees the result of the toss plus not when the coin has been tossed. Let $X_n :=$ "coin tossed on nth toss". The set of observations, in this case, is $\mathcal{O} = \{0, 1\}$ where 0 indicates that the outcome of the toss is "heads" and 1 indicates that the outcome of the toss is "tail" and the set of hidden states is $S = \{m_1, m_2\}$. Suppose that the transition matrix A and the observation matrix B are respectively:

$$A = \begin{pmatrix} 0.7 & 0.3 \\ 0.1 & 0.9 \end{pmatrix}$$

and

$$B = \begin{pmatrix} \frac{2}{3} & \frac{1}{3} \\ \frac{1}{6} & \frac{5}{6} \end{pmatrix}$$

With initial distribution is $\pi = (\pi_1, \pi_2) = \left(\frac{1}{2}, \frac{1}{2}\right)$, and $P\left(X_1 = m_1\right) = \frac{1}{2} = P\left(X_1 = m_2\right)$. The probability that a sequence of 5 throws was $m_1 m_1 m_2 m_1 m_2$ is equal to

$$P\left(X_1 = m_1, X_2 = m_1, X_3 = m_2, X_4 = m_1, X_5 = m_2\right) = \pi_1 a_{11} a_{12} a_{21} a_{12}$$

$$= 0.5 \times 0.7 \times 0.3 \times 0.1 \times 0.3$$

$$= 0.003\,15$$

The probability that the sequence of observations was 0011110 given that the sequence of tosses is $m_1 m_1 m_1 m_2 m_2 m_1 m_1$ is equal to

$$P(0011110 \mid m_1 m_1 m_1 m_2 m_2 m_1 m_1) = \left(\frac{2}{3}\right)^2 \frac{1}{3} \left(\frac{5}{6}\right)^2 \frac{1}{2}\frac{1}{3}$$
$$= 2.2862 \times 10^{-2}$$

and from there we have

$$P(0011110 m_1 m_1 m_1 m_2 m_2 m_1 m_1) = P(0011110 \mid m_1 m_1 m_1 m_2 m_2 m_1 m_1)$$
$$\cdot P(m_1 m_1 m_1 m_2 m_2 m_1 m_1)$$
$$= 2.2862 \times 10^{-2} \times 4.6305 \times 10^{-3}$$
$$= 1.0586 \times 10^{-4}$$

Solution of the decoding problem:

Let $\lambda = (S, O, A, B, \pi)$ be the given model and $\mathcal{O} = v_1 v_2 \ldots v_T$ a sequence of observations. The objective is to determine the "most probable" sequence of hidden states generated by the observation sequence.

There are several ways to interpret the expression "most likely". One possible interpretation is to choose, at each moment, the hidden states that individually are more probable. This approach allows maximizing the expected number of correct hidden states but may even tually generate an unlikely sequence of observations since transition probabilities between hidden states are not taken into account. For this reason, another methodology is required that allows finding the best sequence of hidden states generated by the sequence of observations and that maximizes their posterior distribution. To achieve this goal, the Viterbi algorithm is generally used, which is presented below.

Viterbi Algorithm

For each $t = 1, 2, \ldots, T$. Let $\delta t\,(i)$ be the maximum probability that the model $\lambda = (S, O, A, B, \pi)$ reaches state i at time t, emitting observations $v_1 v_2 \ldots v_t$. This is

$$\delta_t(i) = \max_{\substack{s_1, s_2, \ldots, s_t \\ s_t = i}} P(O_1 = v_1, \ldots, O_t = v_t, X_1 = s_1, \ldots, X_t = s_t)$$

The probability can be calculated recursively, for $t = 2, \ldots, T$ in the following way:

$$\delta_t (i) = \max_{\substack{1 \le j \le n s_1, s_2, \ldots, s_{t-1} \\ s_{t-1}=j}} \max P (O_1 = v_1, \ldots, O_{t-1} = v_{t-1}, X_1 = s_1, \ldots, X_{t-1} = s_{t-1})$$

$$P \left(X_t = s_i \mid X_{t-1} = s_j \right) P \left(O_t = v_t \mid X_t = s_i \right)$$

$$= \max_{1 \le j \le n} \delta_{t-1} (j) a_{ji} b_{it} (O_t)$$

Hence

$$\delta_t (i) = \begin{cases} \pi_i b_{i1} (O_1) & \text{if} \quad t = 1 \\ \max_{1 \le j \le n} \delta_{t-1} (j) a_{ji} b_{it} (O_t) & \text{if } t = 2, 3, \ldots, T \end{cases}$$

The objective is to retrieve the entire sequence of states that maximize the probability of observing the given sequence of observations, then at each iteration the maximizing argument of that probability must be retrieved. To do this, the following variable is defined:

$$\phi_t (i) := \arg \max_{1 \le j \le n} \left[\delta_{t-1} (j) a_{ji} \right]$$

Therefore the algorithm which is proposed is:

(i) Initiation: for each $i = 1, 2, \ldots, N$ it is considered:

$$\delta_1 (i) = P (O_1 = v_1, X_1 = s_i)$$
$$= \pi_i b_{i1} (O_1)$$

(ii) Iteration: for each $t = 2, \ldots, T$ and for each $i = 1, \ldots, N$ are calculated recursively

$$\delta_t (i) = \max_{1 \le j \le n} \delta_{t-1} (j) a_{ji} b_{it} (O_t)$$

and

$$\phi_t (i) = \arg \max_{1 \le j \le n} \left[\delta_{t-1} (j) a_{ji} \right]$$

(iii) Finalization: The value of the maximum probability P^* is calculated among all the N terms $\delta T (i)$ at instant T of the last observation. That is,

$$P^* = \max_{1 \le i \le n} \delta_T (i)$$

and it is determined by:

$$s_T^* := \arg \max_{1 \le i \le N} \phi_T (i)$$

(iv) Reconstruction of optimal sequence:

$$s_t^* = \phi_{t+1} \left(s_{t+1}^* \right)$$

with $t = T - 1, T - 2, \ldots, 1$

Algorithm 3: Viterbi algorithm

Initialization:

> for $i = 1, \ldots, N$ do
> | $\delta_1(i) = P(O_1 = v_1, X_1 = s_i) = \pi_i b_{i1}(O_1)$
> end

Iteration Step:

for $t=2, \ldots, T$ do
> for $i=1, 2, \ldots, N$ do
> > $\delta_t(i) = \max_{1 \leq j \leq n} \delta_{t-1}(j) a_{ji} b_{it}(O_t)$
> >
> > $\phi_t(i) = \arg\max_{1 \leq j \leq n} \left[\delta_{t-1}(j) a_{ji} \right]$
> end
end

Output:

$P^* = \max_{1 \leq i \leq n} \delta_T(i)$

$s_T^* := \arg\max_{1 \leq i \leq N} \phi_T(i)$

Reconstruction of optimal sequence

$s_t^* = \phi_{t+1}\left(s_{t+1}^*\right)$

Solution to the training problem

The Expectation-Maximization algorithm (or Baum-Welch algorithm) is one of the most widely used methods in statistics and machine learning tools to estimate parameters in latent variable models. The algorithm alternates between two steps, the expectation step and the maximization, to compute estimates of the maximum likelihood of the parameters. A weakness of this algorithm is the possibility that the estimates are not global maximums since only the algorithm is guaranteed to converge to the local maxima of the probability function. Initially, there is no knowledge of the parameters best fitting the model λ. Since there is a succession of observations, which is called the training sequence, it will be used to maximize $P(\mathcal{O} \mid \lambda)$. The first thing that is done is to use preselected or randomly chosen probability distributions, and then the most likely observed transitions and symbols are identified. By maximizing the transition probabilities obtained, a new model with a higher probability of generating the observed sequence of symbols is obtained. This process, called the training process, is repeated until the generated model does not differ significantly from the one generated in the previous step. The following shows how the Baum-Welch algorithm works.

Let $\xi_t(i, j)$ be the probability that, given the observation sequence $\mathcal{O}$ and the model λ, the chain lies in time t in state s_i and time $t + 1$ in state s_j. That is,

$$\xi_t(i, j) = P\left(X_t = s_i, \ X_{t+1} = s_j \mid \mathcal{O}, \lambda\right)$$

From the definition of the Forward and Backward probabilities, it can be shown that:

$$\xi_t(i,j) = \frac{\alpha_t(i)\,a_{ij}b_{j(t+1)}\left(O_{t+1}\right)\beta_{t+1}(j)}{P(\mathcal{O}\mid\lambda)}$$

$$= \frac{\alpha_t(i)\,a_{ij}b_{j(t+1)}\left(O_{t+1}\right)\beta_{t+1}(j)}{\sum\limits_{i=1}^{N}\sum\limits_{j=1}^{N}\alpha_t(i)\,a_{ij}b_{j(t+1)}\left(O_{t+1}\right)\beta_{t+1}(j)}$$

where $\alpha_t(i)$ is the probability of "advance", that is, the probability that at time t the chain is in state s_i and that the sequence of observations obtained up to time t, let $v_1, v_2, \ldots, v_t$; a_{ij} be the probability of transition from hidden state s_i to hidden state s_j; $b_{j(t+1)}\left(O_{t+1}\right)$ is the probability that the observation at time $t+1$ given that v_{t+1} the hidden state in the time $t+1$ is s_j. Finally, $\beta_{t+1}(j)$ is the probability of having the partial observation sequence $v_{t+1}v_{t+2}\ldots v_T$ from time instant $(t+1)$ to final time instant T.

Let $\gamma_t(i)$ be the probability of being in state s_i at time t, given the observation sequence and the model λ. That is,

$$\gamma_t(i) = P(X_t = s_i \mid \mathcal{O}, \lambda)$$

$$= \sum_{j=1}^{N}\xi_t(i,j)$$

Adding $\gamma_t(i)$ from $t=1$ to $t=T-1$ gives the expected number of times the process transitions from the state i, and if we add $\xi_t(i,j)$ from $t=1$ to $t=T-1$ we obtain the expected number of transitions from state s_i to state s_j, given the sequence of observations. Making use of the previous development, a method is obtained to achieve the estimation of the parameters A, B and π of the hidden model of Markov. Let

$$\widetilde{\pi}_i := \gamma_1(i)$$

$$\widetilde{a}_{ij} := \frac{\sum\limits_{t=1}^{T-1}\xi_t(i,j)}{\sum\limits_{t=1}^{T-1}\gamma_t(i)}$$

and

$$\widetilde{b}_{kj} := \frac{\sum\limits_{t:O_t=v_k}\gamma_t(i)}{\sum\limits_{t=1}^{T-1}\gamma_t(i)}$$

that is, $\widetilde{\pi}_i$ is the number of times the process is in state s_i at time $t=1$, $\widetilde{a}_{ij}$ is the number of transitions from state s_i to state s_j over the number of transitions from state s_i and $\widetilde{b}_{kj}$ is the number of times the process is in state s_j if the observed symbol is v_k over the number of times the process is in state s_j.

Let $\widetilde{\lambda} = \left(S, \mathcal{O}, \widetilde{A}, \widetilde{B}, \widetilde{\pi}\right)$ with $\widetilde{A} = \left(\widetilde{a}_{ij}\right)$, $\widetilde{B} = \left(\widetilde{b}_{kj}\right)$ and $\widetilde{\pi} = (\widetilde{\pi}_i)$. In [10] Baum showed that if the initial model is $\lambda = (S, \mathcal{O}, A, B, \pi)$ then either $\widetilde{\lambda} = \lambda$ or $P(\mathcal{O}\mid\widetilde{\lambda}) > P(\mathcal{O}\mid\lambda)$, which means that a model $\widetilde{\lambda}$ has been found that is more likely to generate the sequence of observations $\mathcal{O}$. Thus, if the model $\tilde{\lambda}$ is used iteratively instead of λ for re-estimation of the parameters will then improve the probability that the sequence $\mathcal{O}$ will actually be observed from the model. The final result of the re-estimation of the model is called the maximum likelihood estimator of the hidden Markov model. Initial estimates for the EM algorithm can be obtained simply by sampling parameter values at random. However, the algorithm is known to be sensitive with respect to the rate of convergence and the given initial values.

Algorithm 4: Baum-Welch Algorithm

Initialization:

Given A and B

$$\delta_1(i) = P(O_1 = v_1, X_1 = s_i)$$
$$= \pi_i b_{i1}(O_1)$$

Expectation Step:

for $t = 1, \ldots, T$ **do**

    for $i,j = 1, \ldots, N$ **do**

$$\xi_t(i,j) = \frac{\alpha_t(i) a_{ij} b_{j(t+1)}(O_{t+1}) \beta_{t+1}(j)}{\sum_{i=1}^{N} \sum_{j=1}^{N} \alpha_t(i) a_{ij} b_{j(t+1)}(O_{t+1}) \beta_{t+1}(j)}$$

and

$$\gamma_t(i) = P(X_t - s_i \mid \mathcal{O}, \lambda) = \sum_{j=1}^{N} \xi_t(i,j)$$

    end

end

Maximization Step:

for $i=1, \ldots, N$ **do**

    $\tilde{\pi}_i := \gamma_1(i)$

end

for $i,j = 1, \ldots, N$ **do**

$$\tilde{a}_{ij} := \frac{\sum_{t=1}^{T-1} \xi_t(i,j)}{\sum_{t=1}^{T-1} \gamma_t(i)}$$

end

for $k,j = 1, \ldots, N$ **do**

$$\tilde{b}_{kj} := \frac{\sum_{t:O_t = v_k} \gamma_t(i)}{\sum_{t=1}^{T-1} \gamma_t(i)}$$

end

5.2   Application for Animal Behavior

The study of animal movements has gained special attention by researchers and environmental groups in recent years, either to determine high-risk areas, breeding areas or migratory routes, their movements have indicated latent behaviors or interaction with the environment. The information collected to study their movements is generally latitude and longitude with auxiliary information of the environment, however, depending on the localization instrument, incomplete observations or noise in the measurements can be obtained. In this application[1], we will study the hidden movements and behaviors of two animals from their observed trajectories (latitude and longitude). To do so, we will simulate the tracking trajectory of two sharks of 500 steps each, with two different states and with one environmental variable (temperature) using the statistical program R and the moveHMM library.

Before proceeding further, there are additional considerations to those discussed above for modeling the movement with an HMM.

1 Let $X_n :=$ "*Latitude and longitude at the nth moment*". The motion of the animal can then be considered as bivariate time series comprising two important things: the step length and the turning angle at each time point (Fig. 5.1) which starts from the observations obtained.
   The set of observations will be given by $O = \{v_1, v_2, \ldots, v_n\}$, where $v_n = \{$"*Step length at the nth moment*", "*Turning angle at the nth moment*$\}$".
2 The parameters of the step length distribution (e.g., gamma distribution) and the parameters of the turn angle distribution (e.g., Von Mises distribution) are determined by an underlying unobserved state.
3 The finite set of hidden states is $S = \{S_1, S_2\}$ and will be given by approximate classifications of the animal's motion (e.g., more active "transit" vs less active "foraging").

Special considerations for HMM animal behavior.

- **Choice of number of states:** In HMMs, the number of states must be chosen prior to analysis, rather than estimated as part of the model fit. This choice can be difficult, particularly because standard model selection criteria the Akaike information criterion (AIC), and the Bayesian information criterion (BIC) tend to select a large number of states, which are difficult to interpret. AIC behaves this way because adding states greatly increases the flexibility of the model to capture features of the data, but this often comes at the cost of interpretability, for this problem [11] suggested that the number of states should generally be informed by biological experience and model checking, rather than following the usual selection criteria. Normally, with motion data a low number of states is assumed. For example, foraging (low speed, high turn rates) and transiting (high speed, low turn rates).

---

[1] This application is part of the ongoing project work for the student Sara Lucia Acosta Pinzón under guidence of L.B.C.

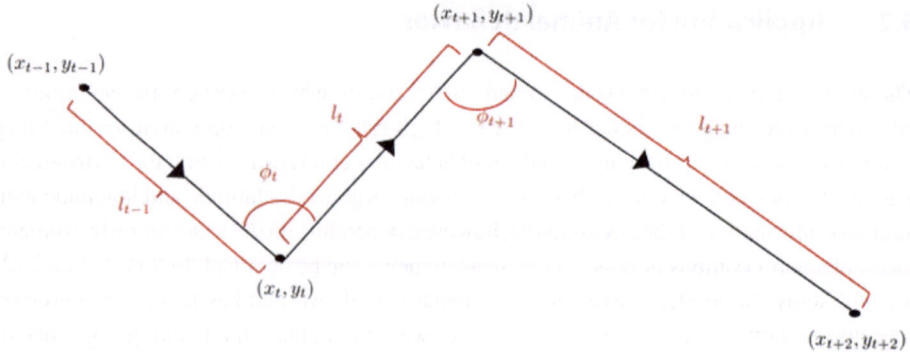

Fig. 5.1 Illustration of step length and turning angle

- **Initial parameters:** Model fitting requires numerical optimization of the likelihood function, which measures the likelihood of the observed data given a set of parameter values. The optimizer requires initial parameter values, from which it begins to explore the parameter space (to find the optimum), this means that the choice of initial parameters must be based on exploratory visualization of the data and can be challenging in many applications, especially for complex models (e.g., with many states) [12].

- **Noisy or irregular data:** The key assumption of HMMs is that data should be observed with negligible measurement error (to obtain reliable step lengths and turning angles) and at regular time intervals. In cases of having irregular measurements, the model cannot be applied directly. A common approach has been to first filter and regularize this data using another model and then apply the HMM approach to the output trajectories. The first step can be applied, for example, with the state space models implemented in the R crawl packages (Johnson et al., 2008). An alternative approach was suggested by [13] implemented in the momentuHMM package, which consists of generating many plausible leads from a state-space model, fitting an HMM to each of them, and then combining the results to obtain estimates that account for uncertainty.

The parameters for this example will be given by the following specifications: the mean for the passage length in the foraging state is 2 km with sd of 2 km and for the transit state it is 10 km with sd of 5 km, additionally, the parameters for the turning angles for the foraging state will be π while for the transit state they will be 0 or close to zero. Finally, the distribution for the passage length will be the gamma distribution and for the turning angles the Von Mises distribution. The simulation is presented below.

```
library(moveHMM)
number_steps  <- 500                           #  steps per animal
stepPar <- c(2,10,2,5,0.2,0.3)                 # step distribution par
anglePar <- c(pi,0,0.5,1)                      # angle distribution par
cov <- data.frame(temp=log(rnorm(1000,35,10))) # Covariable
stepDist <- "gamma"                            # step distribution
angleDist <- "vm"                              # angle distribution
stateNames<-c("Foraging","Transit")           # state names
```

- **Evaluation:** Generate a Hidden Markov Model given a sequence of Observations.

  1 we create the simulation of the two trajectories with 500 steps each, starting from the given initial parameters above mentioned.

```
data <- simData(nbAnimals=2,               # Two trajectories
                nbStates=2,                # Two states
                stepDist=stepDist,         # Step Distribution
                angleDist=angleDist,       # Angle Distribution
                stepPar=stepPar,           # Step parameters distribution
                anglePar=anglePar,         # Angle parameters distribution
                covs = cov,                # Covaribles for each step
                zeroInflation=TRUE,
                obsPerAnimal=number_steps) # Number of steps per animal
```

  2 We can see that the data consists of six columns: a row of IDs (2 different animals), coordinates of (x,y), the step length and turning angle calculated for each pair of coordinates and an auxiliary variable of temperature (Table 5.1). Observations will be given by $O = \{v_1, v_2, \ldots, v_n\}$, where $v_n = \{$"Step length at the $n-$th moment", "Turing angle at the $n-$th moment$\}$"

Table 5.1 Sequence of simulated observations, with calculation of step length and turning angle

ID	x	y	Step	Angle	Temp
1	3.394	−2.888	2.452	−2.369	3.573
1	−2.447	−1.643	1.135	2.642	3.498
1	−1.415	−6.395	3.335	1.111	3.982
1	−1.415	−6.395	3.350	−1.736	3.011

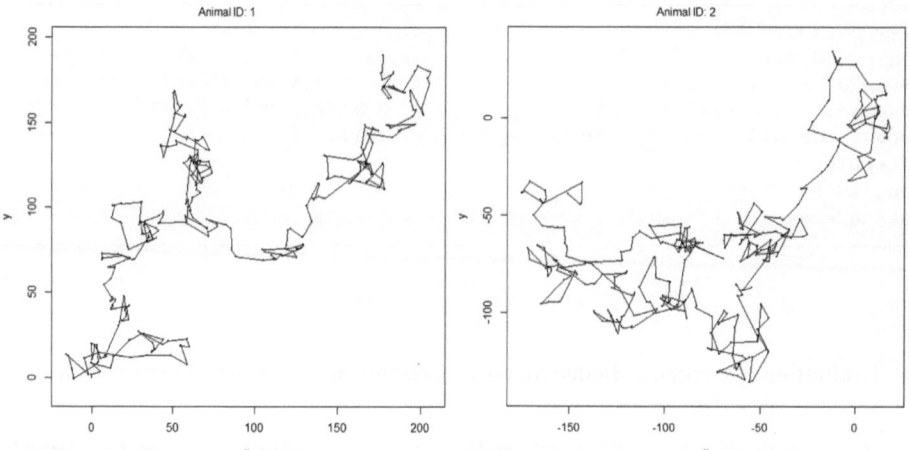

Fig. 5.2 Simulated trajectories

3   We visualize the simulation trajectories in Fig. 5.2, showing the possible route followed by the animal simulating a possible real path.

4   A Hidden Markov Model $\lambda = \{S, O, A, B, \pi\}$ is generated for a Forward/Backward algorithm.

The model explores the parameter space for the mean, standard deviation and zero mass parameters and returns the optimal values for each of the behavioral states. From a biological perspective, the mean passage length of the first state reflecting a foraging behavioral state, this is an order of magnitude smaller than that of state 2 that reflecting a transit behavior. For the flip angle parameter estimates, it is observed that the magnitude of the first state is larger than that of the second state, so effectively the first state is a foraging activity while the next state is a transit state.

```
InitialModel <- fitHMM(data=data,       # Data class moveData or SimData
                       nbStates=2,       # States
                       stepPar0=stepPar, # Initial steps parameters
                       anglePar0=anglePar, # Initial angle parameters
                       formula=~temp)     # Introduce covariable in the model

Step length parameters:

           Foraging  Transit
mean       1.890286 10.3207443
sd         1.870595  4.9522246
zero-mass  0.158997  0.3151645


Turning angle parameters:

                 Foraging Transit
mean           -2.8843556 0.08216547
concentration   0.3717782 1.02739146


Regression coeffs for the transition probabilities:

             For. -> Tra.    Tra. -> For.
intercept   1.37231081       1.01694278
temp        -0.03628231      -0.07863942


Initial distribution:

[1] 1.835883e-06 9.999982e-01
```

- **Decomposition:** The decoding of the behavioral states that most probably generated the sequence of observations for both trajectories is performed with the Viterbi algorithm, from which the distribution of both states is observed as shown in Fig. 5.3 where state one indicates few meters traveled and large turning angles and state two shows large distances traveled, but direction in the movement.

```
states <- viterbi(Model)
states
[1] 2 1 2 2 1 2 1 2 1 2 1 2 1 2 1 2 1 2 1 2 1 2 1 2 1 2 1 2 1 2 1 2 1 2
[35] 1 2 1 2 1 2 1 2 1 2 1 2 1 2 1 2 1 2 1 2 1 2 1 2 1 2 1 2 1 2 1 2 ...
```

The most probable state decoding is obtained, it is possible to observe within the trajectory how the hidden behavioral states of each animal were respectively (Fig. 5.4). From here, it is possible to find species hunting zones, reproduction zones, breeding zones, etc. Depending on the hidden states being observed.

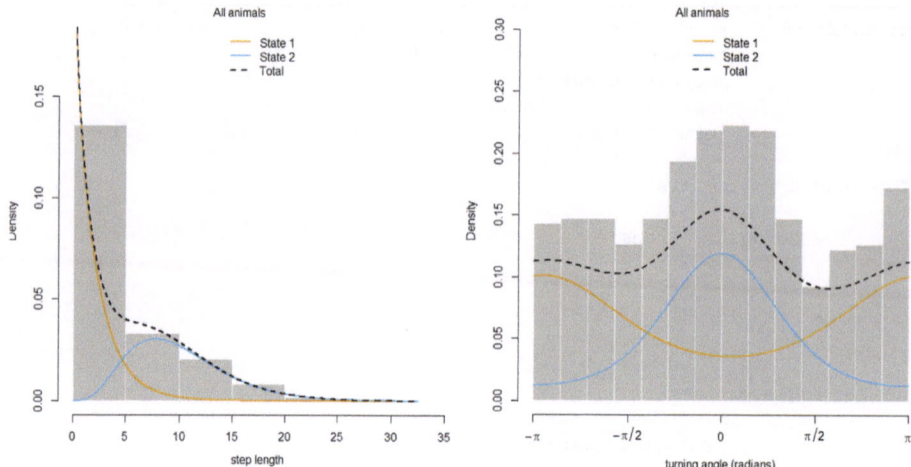

Fig. 5.3 Sequence of decoding states. The right panel shows the state distributions for the state lengths and the left panel shows the distributions for the turning angles, for state 1 "For aging" and state 2 "traveling"

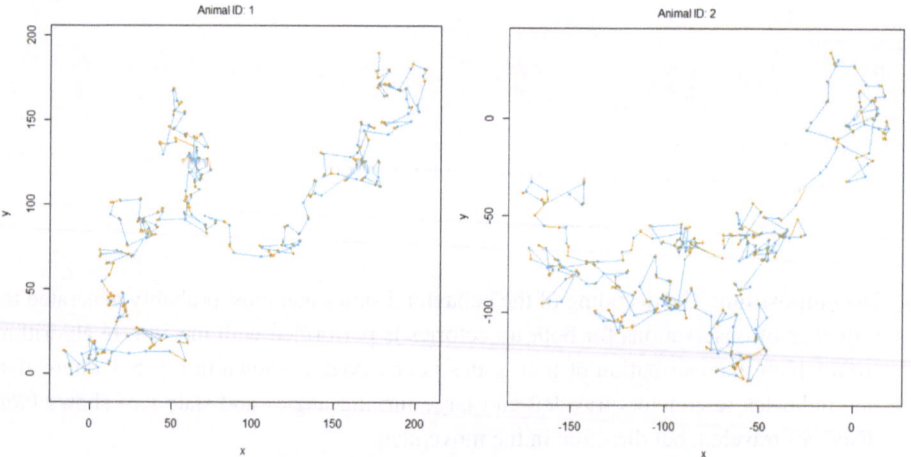

Fig. 5.4 Sequence of decoding states for the two simulated trajectories

Finally, the usefulness and use of the HMMs is emphasized, with them we were able to obtain hidden behaviors (foraging and transient) from the location information of the animals only. This information allows us to determine the proportion of time that the animal spent in one state versus the other (for this example 50.2% in a foraging state and 49.8% in transit), allows us to determine the relationship of their movements with environmental variables, in addition to allowing us to decode the states which indicate the areas where they have these behaviors.

References

1. Baum, L. E., et al. (1972). An inequality and associated maximization technique in statistical estimation for probabilistic functions of Markov processes. *Inequalities, 3*(1), 1–8.
2. Von Hilgers, P., & Langville, A. N. (2006). The five greatest applications of Markov chains. In *Proceedings of the Markov Anniversary Meeting* (pp. 155–158). Boston: Boston Press.
3. Rabiner, L. R. (1989). A tutorial on hidden Markov models and selected applications in speech recognition. *Proceedings of the IEEE, 77*(2), 257–286.
4. Ali, D., Touqir, I., Siddiqui, A. M., Malik, J., & Imran, M. (2022). Face recognition system based on four state hidden Markov model. *IEEE Access, 10*, 74436–74448.
5. Khorasani, A., & Daliri, M. R. (2014). HMM for classification of Parkinson's disease based on the raw gait data. *Journal of Medical Systems, 38*, 1–6.
6. Glennie, R., Adam, T., Leos-Barajas, V., Michelot, T., Photopoulou, T., & McClintock, B. T. (2023). Hidden Markov models: Pitfalls and opportunities in ecology. *Methods in Ecology and Evolution, 14*(1), 43–56.
7. Kosović, I. N., & Fertalj, K. (2014). Discovering the animal movement patterns using hidden Markov model. *International Journal of Computer and Information Technology, 3*, 508–514.
8. Farhadinia, M. S., Michelot, T., Johnson, P. J., Hunter, L. T., & Macdonald, D. W. (2020). Understanding decision making in a food-caching predator using hidden Markov models. *Movement Ecology, 8*, 1–13.
9. Eisner, J. (2002). An interactive spreadsheet for teaching the forward-backward algorithm. In *Proceedings of the ACL-02 Workshop on Effective Tools and Methodologies for Teaching Natural Language Processing and Computational Linguistics* (pp. 10–18).
10. Baum, L. E., Petrie, T., Soules, G., & Weiss, N. (1970). A maximization technique occurring in the statistical analysis of probabilistic functions of Markov chains. *The Annals of Mathematical Statistics, 41*(1), 164–171.
11. Pohle, J., Langrock, R., Van Beest, F. M., & Schmidt, N. M. (2017). Selecting the number of states in hidden Markov models: Pragmatic solutions illustrated using animal movement. *Journal of Agricultural, Biological and Environmental Statistics, 22*, 270–293.
12. Michelot, T., & Langrock, R. A short guide to choosing initial parameter values for the estimation in moveHMM.
13. McClintock, B. T. (2017). Incorporating telemetry error into hidden Markov models of animal movement using multiple imputation, (3), 1537–2693.

Appendix

A

The list of some important distributions:

Discrete distributions				
Distribution	Probability mass function	E(X)	Var(X)	$\varphi_X(u)$
Bernoulli $B(p)$	$P(X = k) = p^k(1-p)^{1-k}, k \in \{0, 1\}$	p	$p(1-p)$	$pe^{iu} + (1-p)$
Binomial $B(n, p)$	$P(X = k) = \binom{n}{k}p^k(1-p)^{n-k}, k \in \{0, 1, 2, \ldots, n\}$	np	$np(1-p)$	$(pe^{iu} + (1-p))^n$
Geometric $G(p)$	$P(X = k) = (1-p)^{k-1}p, k \in \{1, 2, \ldots\}$	$\frac{1}{p}$	$\frac{1-p}{p^2}$	$\frac{pe^{iu}}{1-(1-p)e^{iu}}$
Poisson $P(\lambda)$	$P(X = k) = e^{-\lambda}\frac{\lambda^k}{k!}, k \in \{0, 1, 2, \ldots\}$	λ	λ	$e^{\lambda(e^{iu}-1)}$

Continuous distributions				
Distribution	Density	E(X)	Var(X)	$\varphi_X(u)$
Normal $N(\mu, \sigma^2)$ $-\infty < \mu < \infty, \sigma > 0$	$f(x) = \frac{1}{\sigma\sqrt{2\pi}}e^{-\frac{1}{2}(x-\mu)^2/\sigma^2}$	μ	σ^2	$e^{i\mu t - \frac{1}{2}t^2\sigma^2}$
(Student's) t $t(n), n = 1, 2, ..$	$f(x) = \frac{\Gamma(\frac{n+1}{2})}{\sqrt{\pi n}}d\frac{1}{(1+\frac{x^2}{n})^{(n+1)/2}}$	0	$\frac{n}{n-2}, n > 2$	
Uniform $U(a, b)$	$f(x) = \frac{1}{b-a}, a < x < b$	$\frac{1}{2}(a+b)$	$\frac{1}{12}(b-a)^2$	$\frac{e^{iub}-e^{iua}}{iu(b-a)}$
Exponential Exp(a), $a > 0$	$f(x) = \frac{1}{a}e^{-x/a}, x > 0$	a	a^2	$\frac{1}{1-aiu}$

L. Blanco-Castañeda and V. Arunachalam, *Applied Stochastic Modeling*, Synthesis Lectures on Mathematics & Statistics, https://doi.org/10.1007/978-3-031-31282-3

Index

© The Editor(s) (if applicable) and The Author(s), under exclusive license
to Springer Nature Switzerland AG 2023
L. Blanco-Castañeda and V. Arunachalam, *Applied Stochastic Modeling*, Synthesis
Lectures on Mathematics & Statistics, https://doi.org/10.1007/978-3-031-31282-3